THE LIFE AND WORKS OF J. C. KAPTEYN

THE LIFE AND WORKS
OF J. C. KAPTEYN

by

Henrietta Hertzsprung-Kapteyn

An Annotated Translation with Preface and
Introduction by

E. ROBERT PAUL

*Department of Mathematical Sciences, Dickinson College,
Carlisle, Pennsylvania 17013-2896, U.S.A.*

Reprinted from Space Science Reviews, Volume 64, Nos. 1–2, 1993

Kluwer Academic Publishers

Dordrecht / Boston / London

Library of Congress Cataloging-in-Publication Data

ISBN 0-7923-2603-2

Published by Kluwer Academic Publishers,
P.O. Box 17, 3300 AA Dordrecht, The Netherlands.

Kluwer Academic Publishers incorporates
the publishing programmes of
D. Reidel, Martinus Nijhoff, Dr W. Junk and MTP Press.

Sold and distributed in the U.S.A. and Canada
by Kluwer Academic Publishers,
101 Philip Drive, Norwell, MA 02061, U.S.A.

In all other countries, sold and distributed
by Kluwer Academic Publishers,
P.O. Box 322, 3300 AH Dordrecht, The Netherlands.

Printed on acid-free paper

Printed in Belgium

Dedicated to the memory of
Vic Thoren
– mentor and friend

Robert Paul

TABLE OF CONTENTS

ACKNOWLEDGEMENTS

Without the generous assistance of several archivists, librarians, and astronomers, it would have been considerably more difficult to locate and examine manuscript and source material upon which this annotated translation is based. I am thankful to Dr Adriaan Blaauw of the Kapteyn Astronomical Laboratory who critiqued my translation and thus saved me from certain embarassment; Dr Blaauw also provided me with invaluable information regarding the disposition and early history of Kapteyn's Nachlass. I am grateful to Dr Steven J. Dick senior historian at the United States Naval Observatory for providing comments on the entire book. Thanks are extended to the staff of Wellesley College (Massachusetts) for lending me a copy of this rare book (indeed, only three libraries in the United States possess an original edition), to the staffs of the Kapteyn Astronomical Laboratory and the University of Groningen for allowing me access to original source materials dealing with Kapteyn's life, and to the Huntington Library and the Mount Wilson and Las Campanas Observatories (George Ellery Hale Papers) and Yerkes Observatory (Kapteyn–Parkhurst Papers) for providing access to the extant personal correspondence upon which the original author based much of her work. My appreciation to the American Institute of Physics (Bart Bok Interview) and to the staffs of the United States Naval Observatory Library and the Boyd Lee Spahr Library of Dickinson College for their help in locating a variety of esoteric sources. Finally, I am grateful to my research assistants Ann-Marie Paul, Inez Pot, and Alan Beemer, to Kathleen Paul for proof reading the entire book, and to Dickinson College for a Board of Advisors faculty research grant that enabled me to complete this annotation.

LIST OF ILLUSTRATIONS

PREFACE

Not only does Henrietta Hertzsprung-Kapteyn's biography of her father reflect on Kapteyn the man, but it also sketches – however briefly – on many technical developments. The biography also suffers from – and in many ways is enriched by – the emotional excesses of a loving daughter writing of her very famous father. Consequently, the original biography is understandably excessively weak in its critical assessment of Kapteyn. As a result, it would have been wholly inappropriate to undertake a *critical* evaluation of this biography; the book was simply not written to convey and evaluate the enormously complicated story of Kapteyn's scientific work. For that subject one must turn elsewhere.[1] Although where appropriate I have provided a critical guide through the more technical material, the present translation is fully annotated. Even so what justifies a translation of her biography? I believe there are still two very important reasons.

In the absence of a full-scale, scientific biography, this biography not only provides the best introduction we currently have of the life of a very influential scientist, but an English translation opens-up to a much wider reading public many of the enormously rich contributions, not only of Kapteyn the man but also of the Dutch, to the emergence of astronomy as a major intellectual force in the world. Without access to primary correspondence, however, it is impossible to reconstruct historical and scientific events. In a scientific biography, this material is absolutely necessary. Tragically, however, Kapteyn's *Nachlass* was completely destroyed during the bombing of Rotterdam in May of 1940. Easily the largest collection of Kapteyn's extant correspondence is presently preserved in the George Ellery Hale Microfilm Collection, which contains several hundreds of letters between Kapteyn and Hale.[2] Additional correspondence between Kapteyn and various astronomers, scientists, scholars, and others also exists in a variety of European and American libraries and archives, principally the Kapteyn Astronomical Laboratory at Groningen, the University of Groningen, and the Yerkes Observatory at Williams Bay, Wisconsin.[3]

Perhaps equally important, this biography, both in the original Dutch and in translation, reproduces many biographical and technical details from Kapteyn's correspondence with numerous other scientists and scholars that have otherwise become unavailable as a result of the destruction of Kapteyn's Nachlass. Therefore, access to the Kapteyn biography, particularly in translation, becomes an archival

---

[1] The technical details relating to Kapteyn's scientific work within the larger astronomical and scientific setting may be found in E. Robert Paul, *The Milky Way Galaxy and Statistical Cosmology, 1890–1924* (New York: Cambridge University Press, (1993).

[2] 'George Ellery Hale Microfilm Collection' (Pasadena: Mt. Wilson and Las Campanas Observatories and the Huntington Library). References to this collection are cited throughout as Hale.

[3] References to these archives are cited throughout as K.A.L., Groningen, and Yerkes, respectively.

J. C. KAPTEYN
ZIJN LEVEN EN WERKEN
DOOR H. HERTZSPRUNG-KAPTEYN

P. NOORDHOFF — 1928 — GRONINGEN

Fig. 1.   Title page of Dutch Edition (1928).

treasure for future studies dealing with Kapteyn himself, as well as with the history of both modern and Dutch astronomy, and George Ellery Hale, Mount Wilson, and the rise of American astronomy. Because the disposition of the Kapteyn Nachlass is of considerable importance, let me note the relevant events of its disappearance. In a personal letter to me in May of 1988, Dr Adriaan Blaauw, former director of the Kapteyn Astronomical Laboratory, rehearsed his understanding of the story:

Once I was told by P.J. van Rhijn, Kapteyn's successor at Groningen, that it had been the intention

Fig. 2.   J. C. Kapteyn with Henrietta Hertzsprung-Kapteyn, ca. 1910.

of Willem de Sitter, together with the historian Johan Huizinga, to write the biography of Kapteyn. To this purpose, according to van Rhijn, documents pertaining to Kapteyn's work and life were collected and kept in a chest or trunk, and during the heavy bombing of Rotterdam, which marked the beginning of the war with and occupation by Germany in May 1940, this was destroyed. As you probably know, de Sitter died already about 1935 and Huizinga died during the war. Why the collection of documents should have been in Rotterdam I have no idea. But I do seem to remember that at the time I made my first acquaintance with the Leiden Observatory, a year or so before de Sitter's death (I was allowed by de Sitter to live in the 'assistant's house' at the observatory and observe with the astrograph), there was in the attic old instruments and the library in which was kept a large chest with a collection of documents of an historical nature that puzzled me.[4]

---

[4]   Adriaan Blaauw to E. Robert Paul, 28 May 1988 (in author's personal file).

As a result of the loss of most of Kapteyn's personal correspondence and materials, the writing of a full-scale, reliable, scientific biography may never be possible. Consequently, it has been my purpose to provide an annotated translation. The original biography did not provide any references to the sources – neither private correspondence nor technical literature – upon which the author based her writing. It is clear from the sources quoted, however, that the author had access to Kapteyn's personal, unpublished correspondence. Where possible I have attempted to reconstruct and identify all of these critical sources; those that remain unidentified were most likely lost in the Rotterdam bombings. In addition, not only are all references to peoples and places identified, particularly those central to developments in astronomy, but I have also chosen to clarify all scientific ideas that should equally be understood.

Although I have grouped the original chapters under entirely new headings, I have retained both the original ordering and only slightly altered the original chapter titles themselves, which now appear as subheadings. Finally, the style of the original Dutch biography by Kapteyn's daughter is cultured and colloquial, but exclusively addressed to the non-scholarly community. Although the present translation attempts to convey both the spirit and intent of the original biography and attempts to preserve the many Dutch idiomatic expressions – even though they are more properly understood only in the original – I have above all tried to remain technically close to the scientific and historical material. There is one particular stylistic change that I have made which, it seems to me, renders the biography more readable than in the original: Kapteyn's daughter had written her biography throughout in the third person; I have rendered the present translation in the first person, so that the reader will understand that phrases such as 'father', 'mother', 'grandfather', 'our family' all have as their referent the original author, Kapteyn's daughter Henrietta.

INTRODUCTION
J. C. KAPTEYN AND MODERN ASTRONOMY

Among the most influential astronomers during the period 1900 to 1920 one would certainly include Arthur S. Eddington, George Ellery Hale, Edward Pickering, Hugo von Seeliger – and J. C. Kapteyn. This is not to say there were no others who had not achieved enormous and influential stature. But these few astronomers were among the most influential as astronomy gradually transformed into an *international* community. A Dutchman by birth and a physicist by training, but an internationalist by inclination and an astronomer by choice, Jacobus Cornelius Kapteyn (1851–1922) became a major force, beginning during the first decade of the twentieth century, advocating international cooperation at the deepest and most wide reaching levels. Kapteyn's influence resulted from and subsequently contributed to the 'second golden age of Dutch science'.[1] In the words of Arthur S. Eddington, who S. Chandrasekhar considered 'the most distinguished astrophysicist of his time':

Holland has given many scientific leaders to the world; it is doubtful whether any other nation in proportion to its size can show so fine a record. J. C. Kapteyn was among the most distinguished of its sons – a truly great astronomer.[2]

Although not alone, Holland has never been known for its observing climate; indeed, if anything the overcast, often cloudy and windswept lowlands mitigated virtually all astronomical activities. Consequently, the Europeans had established major observing facilities at Cape Town, Cordoba, and elsewhere, while the Americans were soon to erect world-dominating observatories at Mount Wilson in Pasadena and Mount Hamilton in Northern California. Therefore, working in less than ideal circumstances, Kapteyn initially found his niche with the organization of his world famous Astronomical Laboratory at Groningen, which he used to reduce and analyze astronomical data from other observatories.[3] Thus, for example, in collaboration with Sir David Gill, then English Astronomer Royal at the Cape and the inspiration behind the project, it took Kapteyn nearly thirteen years to reduce

---

[1] Bastiaan Willink, 'Origins of the Second Golden Age of Dutch Science after 1860: Intended and Unintended Consequences of Educational Reform,' *Social Studies of Science*, 21 (1991), 503–26

[2] Eddington, A. S.: 1922, 'Jacobus Cornelius Kapteyn', *The Observatory* **45**, 261; Chandrasekhar, S.: 1983, *Eddington: The Most Distinguished Astrophysicist of His Time*, Cambridge University Press, Cambridge.

[3] In 1922, shortly before Kapteyn's death, the French astronomer Jules Baillaud noted in his opening remarks before the International Astronomical Union in Rome: "The three things that have revolutionized astronomy in the last half century are photography, telescopes, and [Kapteyn's] Laboratory in Groningen"; see pp. 38. In addition to these three developments, the inclusion of spectroscopy into the astronomical corpus should be added to this list.

the data before the *Cape Photographic Durchmusterung*, containing nearly 450 000 stars, was published between 1896 and 1900.

In 1902, Kapteyn discovered the 'two star-streams', which suggested that the Milky Way galactic system was composed of two intermingling and interpenetrating, but distinctly preferential stellar groups moving through one another.[4] Although his most important and lasting empirical achievement, by the late 1920's his compatriot Jan Oort, who became Holland's most celebrated astronomer of the twentieth century and one of the world's most influential scientists, and the Swedish astronomer Bertil Lindblad had determined that star-streaming was really an indication of what astronomers now call 'differential galactic rotation', a phenomenon characteristic of all spiral nebulae.

At the 1904 meetings of the International Congress in St. Louis, Kapteyn proposed his 'Plan of Selected Areas' in which he hoped to enlist observatories world-wide in the collection of raw astronomical data needed to understand the cosmology of the heavens.[5] Some twenty observatories participated in this mammoth project, gradually achieving important results within a decade.

From 1907 onwards until his death in 1922, Kapteyn collaborated directly with the American astronomical giant George Ellery Hale, who, as the founding genius behind the Mount Wilson Observatory, nearly transformed observational astronomy single-handedly. As a result of their relationship and under Kapteyn's recommendations, Hale invited numerous European astronomers to join the Mount Wilson staff. Among those who spent time at the Pasadena facility were Ejnar Hertzsprung, Pieter van Rhijn, H. Ludendorff, Adriaan van Maanen, and Arnold Kohlschütter.

As important as all these programs, discoveries, and developments are, Kapteyn is justifiably known most widely for his studies in cosmology. Indeed, all of Kapteyn's activities can be seen as a reflection of his larger interests in understanding the architecture of the Milky Way Galaxy. The approach Kapteyn perfected in his studies has come to be known as 'statistical astronomy'.

The term 'statistical astronomy' refers to observational studies of stars and galaxies – positions, motions, brightnesses, spectra, and so on – that rely fundamentally on statistical methodologies. Historically, statistical astronomy had been of crucial importance to developments in pre-relativistic cosmology. For during the first several decades of the present century astronomers generally believed that a statistical approach to analyzing the aggregate of stars would eventually lead to an accurate understanding of the architecture of the stellar universe. Although William Herschel (1738–1822) initiated this program late in the eighteenth century,

---

[4] Kapteyn, J. C.: 1905, 'Star-Streaming', *Report of British Association For the Advancement of Science*, section A, 257–65. For a discussion of Kapteyn's discovery of this phenomenon, see J. C. Kapteyn to G. E. Hale, 23 September 1915 (Hale).

[5] Kapteyn, J. C.: 1904, 'Statistical Methods in Stellar Astronomy', *International Congress of Arts and Sciences*, Missouri, St. Louis 4, 396–425. Kapteyn later published the details in his: 1906, *Plan of Selected Areas*, Publications of the Astronomical Laboratory, Groningen. For a discussion of Kapteyn's original ideas, see J. C. Kapteyn to G. E. Hale, 7 May 1905 (Hale).

it was not until a century later that it began to achieve significant promise with the work of Kapteyn and the German Hugo von Seeliger (1849–1924). Roughly from 1890 until the early 1920s, Kapteyn and Seeliger shaped astronomers' cosmological views of the stellar universe.[6]

The basic contributions that Kapteyn (and Seeliger) made to this field of astronomy during the three decades prior to the 'astronomical revolution' of the 1920's were enormous. Their's was not a singular project; many contributed to this research tradition focusing on empirical, conceptual, and methodological problems. In the years immediately following 1900 groups particularly in Holland, Germany, and Sweden developed well-defined centers of research: Critical problems became clearly identified; methodological approaches were developed; dedicated research teams were assembled; formal means for distributing results emerged. Furthermore, each of these defining characteristics deeply reflected national styles of their respective communities.

Although contributions to stellar astronomy were quite different in nature, individual views of the sidereal system not only coincided closely but achieved wide consensus among early twentieth-century astronomers. Understanding the stellar universe using statistical techniques, referred to as the 'sidereal problem' in much of the technical literature, was considered the major (traditional) research program of stellar astronomy during much of Kapteyn's lifetime. Kapteyn culminated his productive and influential career in 1920/1922 with the publication of what the English astrophysicist James Jeans later called the 'Kapteyn Universe', a cosmology that described the nature and architecture of the Milky Way Galaxy.[7]

---

[6] See Paul, E. Robert: 1993, *The Milky Way Galaxy and Statistical Cosmology, 1890–1924*, Cambridge University Press, New York, for a thorough discussion of the developments of statistical cosmology during this period.

[7] Kapteyn, J. C. and van Rhijn, P.: 1920, 'On the distribution of the stars in space especially in the high galactic latitudes', *Mount Wilson Observatory, Contributions*, No. 188, reprinted in *Astrophysical Journal* **52**, 23–38; and Kapteyn, J. C.: 1922, 'First attempt at a theory of the arrangement and motion of the sidereal system', *Astrophysical Journal* **55**, 302–28.

CHAPTER 1

FAMILY LIFE

The Kapteyn Family

For centuries, the teaching profession had run through the Kapteyn family like a red thread. Teachers performed generation after generation; teaching was in their blood, and, if they did not teach at school or an institute, they would teach at home. The first known teacher was Paulus Captijn, who in 1712 was born in Berkenwoude in the Krimpenerwaard of South Holland. In 1756, the Schelenen archives in the town of Heukelom made mention of him as "Paulus Captijn, schoolmaster in this town." Many generations had already been teachers before him. Although the archives mention the names, they do not mention the professions. Following Captijn, the position of schoolmaster was passed on from father to son and, as a result, practiced by many in the family. They became quite famous as teachers, and so it was that in an Amsterdam newspaper of 12 July 1868, during a recommendation of candidates applying to be an English teacher, a Kapteyn was introduced as "one who carries the name of a family in which pedagogic capability is traditional."

Gerrit Jacobus Kapteyn, my grandfather, who owned a well-known boarding school in the town of Barneveld, became a celebrity. Born in 1812 in a place called Bodegraven, where his father had been head of the municipal school, grandfather suffered from the 'English disease', which was noticeable by the large size of his skull. He claimed to have said that on passing the oily bargainers – which is what they used to call the peddlers that moved from village to village – on seeing him offered his mother a household medicine to help him get rid of the disease. As this sort of disease was usually cured naturally, the household medicine worked miraculously well. As was shown later in his life, grandfather was cured completely, because he obtained a physical and mental strength that was a paragon of human nature.

At a very young age grandfather became a pupil-teacher at his father's school, during which he took an unbelievable number of exams. In those days one became a teacher consecutively from the fourth, third, second, and first degrees. (Secondary school did not exist at the time.) Hardly anyone succeeded in becoming a teacher of the first degree. Only two candidates succeeded – one of which was my grandfather. During that same time, he regularly traveled to Leiden to attend lectures on 'Ancient Letters'. As the lectures began at nine o'clock in the morning, he traveled through the night by canal boat. Canal boats allowed the farmers and inhabitants of suburban areas, who came from surrounding villages, to arrive at the big market in Leiden in time to set-up and arrange their stalls for the day.

Consequently, grandfather was always much too early for the lectures. To make good use of this valuable time, he took music lessons in order to learn to play the organ. He had a talent for music and at the age of fourteen played the church organ in Bodegraven, thereby helping out the widow of the former organ-player to whom he donated his salary.

His organ teacher, who thought it was much too early, did not get-up for the lesson, but from his bed yelled down comments to my grandfather. After his lectures at the end of the day, grandfather traveled back to Bodegraven in order to teach and study at night. When he did receive an offer as second in-charge at the boarding school of Mr. Van Wijk in Kampen, however, he was forced to forego his lectures and to continue studying on his own. His daily work was plentiful and tough, yet that did not stop him from studying for his exams every night. And if the fatigue and sleep tended to overpower him, he would put his feet in cold water to stay awake. Although his life was very busy and tiring, he still had enough energy to read the works of Shakespeare, especially during the hours that he was on supervisory duty, which reveals his capability to study and his ability for discipline. He had indomitable energy and perseverance, which later proved essential because he qualified with 'excellence' for his candidate exam in the 'letters' in Leiden.

At the age of twenty-five grandfather married my grandmother, Elisabeth Cornelia Koomans, a twenty-three year-old farmer's daughter from Bodegraven. They settled down in Voorschoten where they opened a boarding school for boys. This task was, however, too much for the strength of the young couple. The youngsters who visited the school most frequently came from the Hague, and they were spoiled, used to luxury, and difficult to guide. After several tough years, they decided to give up the school. Grandfather applied to a boarding school in Barneveld, which included a municipal bonus of 600 guilders and a municipal residence. He received the appointment and they moved to Barneveld with both of their sons; something they never regretted.

The school was established to give boys in Barneveld and the surroundings a greater opportunity to learn than they could achieve in the elementary schools. Through the years the school acquired a very good name and the number of applications grew so large that the boys had to board out in the village itself. Grandfather was not encouraged by these developments, and decided to say farewell to this establishment and build his own boarding school for a limited number of students. He acquired a large acreage and had a stately, massive house called 'Benno', which means 'the strong child', built on it.

Students came from all over the country, and in time it truly became a wonder. Being in a tradition of teachers, grandfather was a born teacher and educator. He was a man of strict justice, love for his work, and of the kind of energy that was never self-serving. His versatility was incredible. The latter was not superficial, however, but rather one of extensive and deep knowledge; two characteristics that rarely go together.

Grandfather spoke three modern languages fluently, and taught Latin and Greek to those students who had to take admission exams for the university. Moreover, he taught geography, history, accounting, and economics, and was known all over the country as a mathematician. His handwriting was like lithography; he never made a rough draft and never made mistakes. He taught himself this art by never handing in anything in which something had been crossed out. The paper on which he would have made a mistake would immediately be torn to pieces; and then he started all over again.

If for some temporary period of time there was no clergyman in Barneveld, grandfather would faultlessly teach his students and children the concepts of the catechism. On Sunday evenings, he regularly held some sort of confirmation class, called 'reading-matter'. If it was not too late, afterwards he would tell stories. This was his greatest gift. He could tell stories in such a fascinating and clear way that he held everyone spellbound. They were mostly mythological tales, as well as robber and ghost stories that were full of the most thrilling adventures. Those who had heard him tell of Cartouche[1], or of the eight-sided farmer, or of the ghost that wanted to be shaved to be free of its curse never forgot them.

The library upstairs in the house was exceptionally comprehensive and rich. Everything read at the time was present: Rousseau, Corneille, Racine, Molière, Goethe, Schiller, Lessing, and many others.[2] The language of conversation was French, which was of great value to the boys, who, at the time, learned to express themselves easily in a foreign language. Grandfather himself was thoroughly acquainted with the French language. Once when he could not remember a required word, however, he learned the entire French dictionary inside out. The boys who knew this tried to catch him off-guard, but they never succeeded. He knew all the words and their meanings.

Grandfather's goal was to turn his students into virtuous and able young men with God's blessing. He was always conscious of having to set the right example; he never relaxed and always paid strict attention to educational manners. A dignified man with a stern countenance, a straight posture, and a skull-cap on his grey hair, grandfather was the type one would call a teacher in the fullest sense of the word.

Just like his own children, the school boys had great respect for him. Grandfather was one of those people who was able to maintain authority by his influence and exemplary behavior. There was never any punishment; it was taken for granted that his guidance would be followed and obeyed by everyone. If he disapproved of something in one of the boy's behavior, he would only need to show it by a certain coolness in his handshake in the morning. Even though he had a huge garden laid out around the school, he realized this was still not enough for the boys; consequently he bought a small forest right outside of Barneveld where simultaneously he had a tent with benches built and a skittle alley laid out. Meals

---

[1] Louis Dominique Cartouche (1693–1721) was a well-known French bandit.
[2] Jean-Jacques Rousseau (1712–1778), Pierre Corneille (1606–1684), Jean Racine (1639–1699), Molière (1622–1673), Johann Wolfgang von Goethe (1749–1832), Friedrich von Schiller (1759–1805), Gotthold Ephraim Lessing (1729–1781).

were eaten early on Sundays with the rest of the day free, after which everyone would go to the forest.

The girls brought bread-bins and another lady from the neighborhood provided tea and milk. The boys enjoyed the freedom by building huts, climbing among the trees by means of rope-ladders, and playing the most delightful games. The boys saw that grandfather lived entirely for the well-being of the school; nothing was too much trouble for him. Throughout their entire lives, they remembered with greatest respect the man who gave them so much to go through life with.

Grandmother also helped just as much to promote the success of the school. During her time spent on the farm, she had learned house-keeping skills on a grand scale. This was fortunate for her, as it was a difficult task. Of all people, however, she was well suited to this task. This woman, my grandmother, who had given birth to fifteen children, managed the household, which at various times numbered up to seventy people, with a steady and skillful hand. Although she was not pretty, she had her father's long eyelids, which my father inherited and which made her seem peculiarly absent-minded. She possessed a natural and calm dignity, was very practical, and had an amazing sense of humor.

Just as at school, there was always the strictest order and punctuality at home. To maintain this order, a strong presence was needed. As a result, grandmother was helped by her two daughters, four maids (who stayed for many years), a lady teacher for the younger children, and two servants, who cared for the enormous garden and made sure the kitchen was always well stocked with vegetables and fruit. Thirteen enormous loaves of bread were baked daily by a village baker who prepared them each evening. There was a real baker's oven. The bread was so excellent that the boys considered it as the best in the whole world and often wrote about it or had it sent to them in later years.

Even the baker himself praised it with the beautifully lofty words: "It seems so heavy, but is so light." Grandmother was famous for her remarkable ability to provide variety with the meals. Her motto was that everything should be at its best, and that nothing was too much trouble to make it such. The mothers of the boys, who heard from their sons about the variety and the exceptional tastiness of the food, often asked her how she could think of all those different kinds of things. At home the boys complained that the food was never quite as good as at school. Even though the mothers had recipes sent from school, it never tasted quite the same.

My grandparents employed two deaf-mute seamstresses who had their hands full with clothes that needed to be mended. Consequently, the children thought that all seamstresses were deaf-mutes. Once when one of the children returned home he was profusely surprised and called out, "Mom, mom, there was a seamstress that could talk!" Altogether, there was more than enough needlework to do for this large family for grandmother and her daughters. While grandfather read out loud, there was constant sewing going on in the living room.

In the evenings, the boys were always in the study room with second-in-commands, where the lessons for the next day were prepared. As a result, the boys saw very little of their parents or of everyday family life. Due to a misplaced sense of justice, there was no distinction made between my grandparent's sons and their students. My father, as one of the sons, suffered from this as well as the lack of warmth and solidarity that the busy family had no time nor occasion for.

My grandparent's home required lots of work, and the days and years seemed to fly by. Besides my grandfather and the five second-in-commands, the eldest sons also started teaching at the school while studying at the same time for their exams at the university.

During the winter and summer months, breakfast, which was eaten at seven o'clock in the morning, was accompanied by readings from the Bible because my grandparents were very pious. Lessons started at half-past-seven; the other lessons started at eight o'clock. Although assisted by my grandfather, the second-in-commands had to prepare themselves thoroughly for their own lessons. Even his sons, who taught Latin, Greek, and algebra, had to be thoroughly prepared, because grandfather checked them and listened often.

A select group of the boys was prepared for higher levels of education, while the remainder were prepared for everyday, more practical life. They all reached their goals excellently equipped with knowledge, as well as a sense of obligation and punctuality. Throughout their education they had seen people work hard and so they themselves had learned to work – an invaluable lesson for life.

In the meantime, my grandmother indefatigably kept up her difficult task with an even-tempered calm and order. Her daily work was never done:

> She is decorated with strength and delight
> She laughs with every coming day
> Her mouth opens in wisdom
> A pious instruction is on her tongue.
> She checks the ways of her house mates
> And does not eat bread slowly.
> Her sons appear and call her happy
> Her husband praises her too.
> Many daughters claim themselves mother-hens,
> But you beat them all.
> Grace is deceptive, beauty is idle.
> A woman who praises Jehovah should be praised.

The Spanish mystic Fray Luis de Granada[3] composed these old lines four centuries ago for a woman just like grandmother. Nobody has ever sung her praise more beautifully.

Everyday husband and wife would walk the garden grounds arm in arm. They discussed the many questions and interests pertaining to the school and the house-

---

[3] F.L. de Granada (1504–1588) was a Spanish religious (Dominican) thinker whose writings are ascetic rather than mystical.

hold. This daily hour, during which they could take their minds off work and enjoy a moment of rest and the outside, was deeply appreciated. The cooperation and socializing between these two people was beautiful. Indeed, they did not live a tough life for nothing, because in the evening of their lives stood a young and strong generation ready to go into the world well-equipped.

Siblings

Altogether there were fifteen children, of which the youngest was born after my grandparent's silver-wedding. One son died at a very young age due to a sunstroke, but the others grew to maturity. Without exception, all were physically and mentally healthy, something that rarely happens in big families. Perhaps this was not so strange realizing the exceptionally healthy parents. Both my grandparents were strong and had an iron will and a strong mentality. One needs a big hand and a strong head to prepare forty boys for life. Unquestionably, both possessed these two qualities. The children were worthy scions of these spirited parents, who were honest and righteous, energetic and intelligent. The lives of their children witness to that. During their youth, there were no family ties between the children. They considered themselves students from the same school, preceded only by the example of their parents. It is therefore characteristic that later in life they remembered little of each others' youth. They passed each other by, and as they possessed more intellect than heart, more righteousness than tenderness, they did not feel the want for either. One of the children, my father, possessed tender feelings and suffered the sorrow of a child that feels alone and misunderstood. In his case, this cast a shadow on him during his entire life.

All the children, however, were successful in life. My grandparents gave the boys the opportunity to study at the University. Adriaan, the third son, wanted to be a sailor, but grandfather strenuously objected. Uncle Adriaan refused to give up the idea, however, and was not intimidated by his father who threatened him with forever foregoing another chance for an academic education. When Uncle Adriaan came back after a year at sea, now thoroughly sobered and disillusioned, he wanted to become an engineer. The only way open to him, however, was to start as a common factory-worker. Slowly he climbed the ladder of success and reached the office of head-engineer with the national railways.

Grandfather considered Albert, the fifth son, unsuitable for studying. Consequently, he apprenticed Uncle Albert to a blacksmith, and afterwards as a voluntary worker at the Rhyns railway factory in Utrecht. On recommendation from a friend who had seen him work, grandfather gave Albert another opportunity to study in Liége under condition that he would take his entrance examinations within three months. With all the will-power and strength of mind he possessed, Uncle Albert concentrated on his studies, which, for a boy who had not received more than an elementary education, was an enormous exertion. But he succeeded and obtained first in the examinations, after which he studied at the technical high school (now

Technical University at Delft) for three years and developed a brilliant career. Five other boys also studied in Utrecht, none of whom ever failed an exam. Unfortunately, there was not enough money to pay for the education of the two youngest sons. At ages fifteen and sixteen they went to a trade office in Amsterdam.

The daughters also took many exams as well and were distinguished from others because of their intelligence, perseverance, and a feeling of obligation. All the children excelled in whatever they attempted: doctor, engineer, scientist, teacher, nurse, business man. They were all superior in their work and they delighted in their unlimited confidence as the name 'Kapteyn' had a classy touch to which each, in his or her own way, had contributed a little. They were proud of each other, and after the death of their father, they made it a habit to hold a family reunion every year with their elderly, respectable mother as the center of interest. After her death, they continued the reunions.

Such was the environment in which my father grew-up and matured. Born in 1851 as the ninth child, he was named Jacobus Cornelius after the sister of his father. This Aunt Ko, to whom he also owed the abbreviation of his name, which he hated, was a very talented woman. She possessed considerable intelligence and a great heart. Although she lived economically from the money she made running an all-girls boarding school, she paid for the education of her second brother who wanted to become a doctor. She had a noble face with a high forehead and clear, intelligent eyes, and a calm, dignified bearing and strong, white hands. A stout, energetic woman, she played her part in the forming of her godchild, my father.

Early Years

Although he was never severely ill, my father was a fragile child (and the only child) who caused his parents to worry about his health. He was slender and pale and had a high-pitched voice. Considering his bearing and facial expression, he gave an absent-minded and introverted impression. This view was reinforced by his long upper eyelids.

Contemporaries don't remember much of him except that he was very smart and zoomed through elementary school in an incredibly short time, so that soon there was nothing there for him to learn anymore. They also remember that he was a slim, pale boy with a long, skinny neck, an absent-minded look in his eyes, and a slovenly appearance. There were simply too many other things on his mind to think about his clothes. His mind was constantly occupied. He lay awake for hours mentally calculating or thinking about a problem. He was constantly in deep thought and became so self-absorbed that he virtually forgot everything about him seeing nothing but the idea at the center of his attention. As a result, for example, while on a school trip he walked straight into the canal surrounding the 'Schaffelaer' not coming to his senses until he felt the ground disappear below his feet. He was teased daily by the other boys. On another occasion when the school had organized an excursion in order to visit the monument at Nellestein, in the neighborhood of Doorn (near Utrecht), he was so engrossed in the contemplation of the details of the monument that, while backing up to get a better view, he forgot he was standing on a small hill and fell down backwards, fortunately without serious consequences.

He was an animal lover who kept birds and rabbits and who developed the ability to recognize lots of birds by their song. The feeding of his animals cost him a good deal of worry, as the children did not receive pocket money. He considered selling pretzels at school, which, though a lot of trouble, brought him a little money. For two cents a week he delivered the daily newspaper to the local clergyman, and his mother also paid him some additional money for searching for eggs. In this way, he managed to work enough to make the needed money. He persevered in whatever he attempted because he did not let disappointments and failures discourage him and because he knew how to surmount all challenges. That's why his canary stock farm was such a great success, as every breeder knows how much care and dedication is required.

Once he had caught a squirrel, which he placed in the tower chamber and took loving care of it. Although he brought it the most delicious snacks, the animal started to languish sitting quietly with a sad countenance in a corner of the room. The boy's tender heart suffered until he suddenly realized: the animal needed water. Although others had told him that neither squirrels nor rabbits need water, he immediately fetched a bowl from which the squirrel drank eagerly. After finishing the last drop, it stretched and one, two, three it was up in the curtains. To him this was a happy moment that he never forgot.

Another time when my father was a boy he had caught an owl. Because he had heard that these animals could not see during daylight, he wanted to find out more about it. He criss-crossed lots of strings through a room and set the owl free. The owl evaded all the strings, diving beneath them and fluttering over them without ever touching one. As a result, he concluded that the day-light blindness was just a fable. In this way, life taught himself a lesson of invaluable worth: the importance of his own investigation.

When my father was fourteen years old, his four year-old sister Bertamie return-ing from England brought a star-map with her that they studied together. With a telescope placed in the garden, the stars were easily located. With great care and accuracy, my father made a star-map on which he glued the biggest stars in silver (cut out from color paper) which he then painted. Luckily this work survives and it shows us how even at fourteen years of age he already possessed the same love and accuracy for his work as he developed during his scientific career. He became excited about this, and from this time forward he was serious about his study of astronomy. When his father noticed his seriousness about studying the stars, he bought him a big telescope, which was set up in the tower chamber, through which he diligently observed the stars. Although he had a keen mind, while still a youngster my father had no idea what he would eventually become.

Nobody could argue with him, however. There was a small shop in Barneveld

– at the time every inhabitant of Barneveld would remember the shop of Jan and Maatje den Oudsten – where numerous people assembled to discuss politics and other local events. Laughing, my father could out-talk most of the elders many of whom weren't exactly dumb either. Although he was not in the least aware of it, others felt my father's superiority in many aspects of life. He remained like this through his entire life.

My father's eldest brother, uncle Hubert, who while a student came home for vacations, played chess with his father, my grandfather, every night. Ko (my father's nickname), who was about ten years old at the time, would stand and watch. Once Uncle Hubert said: "Do you want to play a game with me too, little boy?" Grandfather smiled cunningly as he knew the "little boy's" abilities. The game began. Hubert the student was beat in the shortest time. This irritated Uncle Hubert and he proposed a rematch. Hubert lost this one too, which infuriated him even further, and he said, "If I lose once more, I will never play you again." For the third time he lost and he refused ever again to play his little, talented brother again.

Another time my father asked grandfather to help him with a mathematics problem he did not understand. As his father was busy, a friend of my Uncle Hubert said, "Come here and I'll help you." But, oh woe, when he saw the problem – it went far beyond his comprehension.

My father's brothers could not accept that they had to lose both game and conversation to their younger brother, so they let him be. My father found no friendship with the other schoolboys either. Nobody worried much about him, because he was quiet and shy and with every passing day he became more introverted. Yet, he did not like to be superior, because he was sociable to the core, not only mentally but in everything that pertained to love and warmth. His grandparents too had little time for him. Only when he was ill did he see his mother loving and worried. She would sit by his bedside and call him "my little man" and he would temporarily bask in this beneficence. But as he got well this passed and her motherliness once again was buried under the worries of the day. His gradual isolation within his big family became a deep sorrow whose scars still remain.

Although his family probably never noticed this, as they differed from him distinctly; life, rich and full as not many have known, compensated him. One can find the key to his future in this lack of familial love and his unsatisfied yearning for it throughout his youth. Even though this explains the joy he felt at each overture of love, he was always surprised that so many people loved him. Still he responded quickly to anyone who approached him with warmth and friendliness. Father was not critical and analytical in his dealings with people, but he accepted, responded, and lavished upon many the wealth from his warm heart. Later, while in America, he found many kindred spirits. He found Americans to be less critical and uncomplicated. They did not search or wait for things that were not there, but instead they offered friendship and cordiality like children who tend to throw their arms around each other's shoulders even in the early stages of a friendship. It is really blissful to be able to trust and admire as children do. Would one always want to be deceitful? Surely such a person would not speak from experience.

Since his youth father enjoyed female companionship, perhaps because his male contemporaries were unable to better him in anything, and, consequently, they left him alone. He was extremely self-conscious, however, of his timidity and slovenly appearance. Although he never thought about his clothes and appearance, in the company of young girls these things bothered him, and silently he admired the facility with which his brothers and the other boys from school were accepted by the members of the opposite sex. For example, once there was a gymnastics performance in the big hall at school to which many boys and girls had been invited. The boys did all sorts of clever stunts and were applauded; they talked with ease to the young ladies and gave the impression of an enviable grace and ease to the timid boy. All-the-while, father stood in a corner feeling timid and awkward. Then he got an idea: he would prove he was a man too and lugged a huge pole on his shoulders, which was actually much too heavy for him. With utter exertion he carried it through the hall. It was only later that he came to understand why this stunt did not get him any success. Disappointed, he slid back into his loneliness because once again he was misunderstood.

He was a sturdy and resolute child. Once when father accidentally cut himself with an ax, while keeping the whole thing quiet from his mother, he silently went to the pharmacist for a sticking-plaster. Eventually, however, she found the holes in his pants and the blood stains on his clothes, after which she cared for him. This bravery with physical suffering remained characteristic of him throughout his life.

At the age of thirteen, while playing in the church tower, he had tied down the rope of the bells. During the night he was awaken by the sound of the rain against his window and he became afraid that the rope would shrink. His imagination created fearful visions: the clock would be jerked loose and become severely damaged and he would be to blame. Thus he got up in the middle of the night, walked through the sleeping village, and untied the rope in the tower. He never dared tell his parents about it.

During this time of his youth, he seriously thought about a variety of reforms in order to make industry healthy. He did not exactly know how he would accomplish this, but he started very childishly with a campaign against top hats, which he thought foolish and unworthy. When he told his mother, who was very practical, about his vague world-reforming plans, she responded: "And you want to reform the world?" He answered earnestly: "Yes mother, if I, with my weak ability, could make the smallest difference, I would be content." His mother looked at him in surprise and was somewhat impressed with his will-power and the youthful ability of her son, whom she really knew very little. Later he wore the high top himself and, although he did not reform the world, he shared his imagination with others in order to develop his reconcilable and idealistic vision with various groups of people.

This quiet boy had a pious heart and a childish belief in a God as a helper in times of trouble. As he grew older, however, he lost this belief, because he came to believe that it made one rigid and dogmatic as it had with his parents. Because he found no warmth and inspiration in them, religious words simply remained words. Consequently, he remained unsatisfied with the religion of his youth.

Each generation seems to battle the older one. At this period of time, he felt that the conflict concerned the orthodox beliefs of religion, such as obligatory church visits, bible reading, praying, and the traditional expressions of a belief that could not possibly have true worth to a young, searching spirit. Those phrases that had obtained their unassailable respect from tradition had lost their glamour for my father. My father felt that a strong and deep heart did not need dogmatic prescriptions for the soul. Rather, humans create their own religion and their own ideals according to which one wants to live. This identity is the divinity of man.

And so he fought the battle, the only way to reach the top. The supreme freedom of the soul is to find laws oneself and choose the way one has to follow. And with one's own identity there also eventually develops respect for the freedom of others, something that must also be unassailable and holy.

So it was that every prejudice based on rank or class became foreign to him. This was not because of a conviction that grew into a realization. Rather, it was perfectly natural and instinctive, as my father's spirit rose far above all pettiness. Once when he hadn't arrived home, everyone began looking for him in the village – they found him sitting in the tabernacle. To him man was sovereign and his loftiness went beyond coincidence and circumstances concerning his position, environment, or religion. Already at a young age, this train of thought had rooted itself deeply within him. All was healthy with him of course; the strange distortions of the civilized industry did not affect him. These things stayed with him his entire life. Later, after he obtained success and position, his ideas did not change. He greeted old maids as politely as he did the highest of personalities. His clerks loved to remember how he always removed his hat for them first. He possessed a truly democratic spirit, and his whole life was built around this humble simplicity.

At the age of sixteen, he took the entrance exam for the University of Utrecht, but his father judged him still too young to become a student. Consequently, he stayed home for another year. As he had finished school, he had a lot of free time, and often attended class lectures just for fun. He took walks with the school boys, and worked some on his own time, but he concentrated so well that, upon entering the university, he had already gone through the better part of the first year's curriculum.

He was gifted in all subjects, especially in algebra where he followed his own methods and ideas. There was a difficult algebraic book entitled *Cours d'analyse de l'Ecole polytechnique* (1857–59) by C.-F. Sturm (1803–1855) that all his older brothers had studied. He only studied the elaborate subject table and not the book itself; consequently, he solved problems his own way. When he had to take a preliminary examination at Utrecht, the mathematics professor Guniois disapproved of his methods of solution, because they were entirely different than those of the regular program (which Kapteyn later admitted were indeed better).

During the year home before entering the university, he read a great deal from his father's comprehensive library. If he was found missing, one could always find him in the library absorbed in a book. Throughout his life, he cited works that he had read as a youth, such as those of Goethe, Rousseau, and Lessing. He especially loved Lessing's *Nathan der Weise* (1779), which brought him to tears while on his deathbed; Haberling's 'Der König von Zion', whose stately hexameters he could cite so easily; the well known 'Der machtige Knipperdolling!'[4]; and Alphonse Daudet's *Tartarin de Tarascon* (1880), the book with the most delicious boasting ever thought out. He laughed as spontaneously about it on his deathbed as he had when he had read and enjoyed it as a boy, because up until his death he remained young at heart with a boy's happiness and flexibility.

Student Years

My father's parents wanted him to study theology. Because none of their other sons had wanted to do so, they hoped that this quiet, pious child would fulfill their greatest wish and become a clergyman. They knew nothing of the boy's inner thoughts; nothing of the change that had been taking place inside him. They did not know that his spirit had followed its own independent ways for a long time and that he could not take the path they would like to have chosen. As with his other brothers, he knew that he was more suitable for the sciences. He dared not tell his strict father, however. So with another friend from school, who was also afraid of the same thing, the two of them made an agreement to tell their fathers, on the same day, of their true wishes. Supported by the fact that there was another going through the same situation, they found the courage to speak. It turned out to be easier than it had seemed as grandfather, though disappointed in his own dream, realized all too well that theology did not attract his son. So he also consented to allow this son, as with his older brothers, to study mathematics and physics. In 1868 father was admitted to the University of Utrecht where two of his brothers also studied. Still the brothers did not see much of one another, as they lived independently of each other as they had done at home. Just like all the Kapteyns, father acquired the nickname 'Dux', a name they all valued highly. At the university father became the student of the meteorologist Professor Buys Ballot[5] and the mathematical Professor Cornelius Grinwis.[6] The course of study was not difficult for father, because in his free time at home he had already worked ahead a great deal.

Thus he was able to take his studies calmly and lightly, and still enjoy life. Because he was the only one registered in the Philosophical Faculty at the time,

[4] Bernhard Knipperdolling was a sixteenth-century Anabaptist leader.
[5] C. H. D. Buys Ballot (1817–1890) was a Dutch meteorologist and one of the chief organizers of the science of meteorology.
[6] C. H. C Grinwis (1831–1899) was a professor at the University of Utrecht (1867) until his death.

he joined a judicial club from which he gained a great deal of happiness and free thinking that positively complemented his life of duty and obeyance. This change in collegial interaction did not hurt him, because in those days it was normal that the students played around some, something that father enjoyed. When this more leisurely life started affecting him personally, however, he had the strength to quit.

His greatest happiness, however, came with his friendship with Andree Wiltens. Andree was immediately affected by my father's personality and greatness of character by responding warmly. The heart of this lovely boy was warmed by this and he blossomed through a love and friendship as he had never known, yet had always longed for in his life. Andree was the son of an East Indian civil servant, whose three sons studied in Utrecht and who all belonged to prominent organizations. He was a noble and fine human being and possessed a great but rare quality: he could admire and love with all the warmth in his great heart. Consequently, Andree acquired a great admiration and friendship for his talented friend, which had deep meaning for both of them during their entire lives. What meant most to Andree was love.

Father did not only study, as there was a lot of socializing and joy between friends, which Dutch student life can offer. He fit in very well with his joyful and sociable personality. His humor and sociability, the happy side of his being that had been hidden for so long, began to develop itself now. He could also be very cutting, however, which hurt all the more as his criticism was educated and strictly logical. As a result, he made enemies, which bothered him because he could not stand enmity and would rather be friends with everyone. Later in life he lost this cutting tendency more and more with his harmonious demeanor. Many could not even believe that he had ever been so sharp.

At the university, the first examination was moderate. In those days, examinations were open to the public and so many people had come that the room was completely filled. Grandfather was present too and in close vicinity where the younger Kapteyn was sitting. It did not go too smoothly and people became restless. The professor asked father something about a subject he had not studied, and father did not know the answer. The examiner continued questioning on the same subject, though he realized the boy did not have a clue. My father's logical approach became irritated by this and finally he said aloud, and very determined: "Professor, I know absolutely nothing about it!" The room became deadly silent, and he saw his own father grow pale. The Professor then switched to another subject, which father did know a great deal about and so, finally, he did get his diploma.

Afterwards, grandfather accompanied the two boys to their rooms. The tone was happy and lively and, to everyone's surprise, the usually formal father snuggled into a chair, propped his legs up on the window sill, and tipped the ash from his cigar out the window. Was this their proper, formal father? "Yes, boys, the upbringing is over, and now I don't have to set the example anymore." He laughed and enjoyed their surprise at his behavior. In the evening, grandfather joined his son's friends in the student bar to celebrate the event. Surrounded by those present, the students unanimously called to him: "The old Dux has to narrate!" Since they knew of his talent, so the old Dux told stories until everyone forgot the time.

Reciting was hardest for my father. He had a high and shrill voice, his intonation was imperfect, and his recitation restless. Once, during a student recitation night, he had to recite Schiller's poem *Der Taucher* (1793). Someone in the audience voiced: "I didn't think it possible for anyone to recite so poorly." Instead of insulting him, this cutting remark inspired father to practice. Several years later when he had to deliver his inaugural address as a newly appointed professor in the University of Groningen, he committed all his efforts for the preparation. Thirty times he recited his speech to two of his sisters, who sat in the far corners of the room and patiently served as an audience. As in everything he undertook, he also became successful in this. Later, when he had to speak in public, his speech was calm and clear, so that everyone understood and enjoyed it.

The Family Kalshoven

During his last year as a student at the university in Utrecht he made the acquaintance of the Family Kalshoven who were completely different from the Family Kapteyn. One could call them antipodes. Neither of the two Kalshoven daughters were burdened by any responsibilities; there was no studying and working going on, everything was cheerful and cozy, one enjoyed life in a simple and artless way. Merriment, music, and courtships, all those delicious things present in youth filled the atmosphere. Mrs. Kalshoven was a woman of fine culture and spirit. As a young girl, she had married the wood buyer Kalshoven, who was, at the time, a widower with five children. She then gave birth to three more children, two girls and a boy. It had been a heavy task for the lively, young girl, who came from a calm life in Harderwijk, where her father was a senior master at the gymnasium, and who lived in a large, old-fashioned house on the boulevard in Amsterdam. Business was not flourishing, and the family decided to move to Abcoude where life was simpler and cheaper. Mrs. Kalshoven was a tough lady, who knew how to adjust herself to the circumstances as they presented themselves. The years that followed were happier after the difficult time in Amsterdam. The stepchildren of Mrs. Kalshoven grew up and left home, and after her husband's death, Mrs. Kalshoven moved to Utrecht with her own children, where she had to live economically and simple.

As she grew older, she had trouble walking and became dependent on her chair. She resigned herself to this trial, but there were also days that it got too much for her and she suffered greatly from her handicap. She could wail and complain on occasion only to be so gay and happy on another visit that both daughters would look at each other and think the problem not too serious. In their youthfulness, they could not even imagine what it meant to be an invalid or even what invalidity meant.

Their mother was surrounded by an unconditional love especially from her oldest daughter Elise, who was the quietest and most serious of the two girls. Calmly Elise assumed the household tasks. Her greatest joy was to play anything she could think of on the piano. She had great musical talent that she developed by herself, because the expense of lessons was out of the question in this simple family. Marie, the younger sister, had a beautiful voice, and the two girls would play and sing together for hours. Marie would think of a melody and start singing. Elise then followed her on the piano, which was not a problem as she was an artist by the grace of God. They did not develop these skills as a result of good lessons and serious studying. Maybe that was unnecessary, however, because were not the kind of people who studied and worked rigid? Rather the sisters lived like birds in a forest who sing and cheerfully play until their heart's content when the sun shines and the sky is blue. Throughout their entire lives music brought them happiness. Their mother too had an impulsive personality and lived one day at a time. She could not resist the temptation of a beautiful summer day and she would call: "Children, the sun is shining and it is so wonderful outside. Let's take a trip in a carriage!" Their thriftiness and worries about the future were momentarily forgotten and one cheered while happily going outside.

Father loved coming here where he became a frequent guest. Here he found a carefree spirit that he had never known at home. Rather, home had always been consumed by duty and work, while here one did not turn-on the evening lights on summer nights in order to hover over books; dusk grew slowly. Young girls' voices sang and danced in front of the open window, and all sorts of jokes were made. Even though it was a bundle of happiness, the seriousness of the eldest girl's personality did not escape father. This small, dark complected girl with the heavy braids and the striking brown eyes had a dignified confidence, a caring soul, and an absolute lack of awareness of her own attractiveness. Because they had never gone to school and learned only little from a governess, they still possessed their own individual originality that contained a certain charm that everyone took to readily.

One evening when father was visiting the Kalshoven family, the door was opened by Elise, who radiated happiness and joy. Affected, he asked, "What's happened?" "Oh, I'm so happy, my brother from India came home," was her reply. Then suddenly it became clear to him that he wanted no other for his wife but this girl, who personified happiness and who didn't have to be made happy like so many others. (He had thought a lot about life's values and what was important and what wasn't.) Father's uniquely healthy and genuine spirit knew, at a young age, how to understand and differentiate, while for others life was still a chaos of impulses and confusion of dreams and seductions. He saved the picture of this radiant young girl in his heart until his time would come.

CHAPTER 2

THE BEGINNING YEARS

Leiden

So passed his wonderful student days. My father had worked hard, the doctoral exam was done cum laude, and on 24 January 1875 he was conferred his doctor's degree in mathematics and physics based on his thesis, 'Investigation of the Vibration of a Membrane'. For the first time, father had to deal with the issue of what he wanted to become. His father had already decided that his son, my father, should become a teacher. But my father wanted nothing to do with that. 'Teacher at a school?' Although his father was very angry at this rejection, my father felt that that was neither his calling nor what suited him. The teaching profession was something that the Kapteyns had ingrained in us children since birth, but it was time that the younger Kapteyn realized that he would now have to find his own way through life.

But the young man held his ground; he wished for a scientific career and went to find one. He had heard of an observatory in China, and, after making it known to the directors that he wished to join them, he was offered the post in Peking as the observatory director. Although he enthusiastically accepted the position, nothing came of it since the observatory never became a reality. Coincidentally, at the same time an observatory post became available at the observatory at Leiden. After inquiry, he solicited the directors and got the appointment. His decision once again took him to the love of his youth, so he tackled this challenge. He stayed two years at the observatory in Leiden, working in those years with all his youthful strength, fortitude, and determination. As he expressed later, "In my whole life, I never worked so hard as I did at that time." He allowed himself little rest, observing, so long as the weather permitted, late into the night. In the mornings, he was earlier to work than the others. He lived so intensely for this science that in those two short years numerous plans formed in his head that would never leave him and upon which he would base his work for the rest of his life.

Classical astronomy focused primarily on the planetary bodies of the Solar System. Known stars served only as reference points for the observation of planets and "as the hands of a great time piece from which sailors can read and which the moon is the big hand." At the end of the eighteenth century, William Herschel[1] began to study the sidereal system. He viewed the heavens as one whole, as one organism, and developed his own methods of discovery and observations in order to uncover its structure and organization. Following Herschel, there were many astronomers in the nineteenth century who focused on this problem without bringing the problem any closer to solution. In order to uncover this mystery of the heavens, father felt strongly drawn to develop new techniques and to explore where no one had ever gone before. He sought not for selfish ends, but to serve. In him glowed true love and complete submission to the guiding star of Science, to which he brought sacrifice, patience, and perseverance. And above all he was a patient worker, with an iron-will to succeed and a shining optimism that won over all possibilities. The saying, "The child is father of the man," applies to my father. Indeed the child had raised the man, and self-discipline and perseverance shaped him from his youth for this heavy task that life had intended for him.

He learned both astronomical theory as well as the ability to handle astronomical instruments. Because of his care in systematically eliminating all mistakes, his observations were of an amazing accuracy. He was not the only perceptive astronomer though. As Arthur S. Eddington[2] later put it: "There is a region lying between purely observational and purely theoretical astronomy in which Kapteyn was unrivalled."[3] This was due to his ability to achieve great accuracy, to sustain searching criticism, and to his wonderful native intuition.

Although he laid the ground-work for his scientific career in Leiden, he did not waste chances to enjoy himself. He made good friends who remained close throughout his entire life. He shared meals with other young men, all hard workers at the beginning of their careers, who were also as enthusiastic as he, a round table of courage and strength. Among others, there were the scientists W. Burck[4] and A. Hubrecht[5], and Hoek who would later become the leader of the North Sea Fishermen from Den Helder – all men of earnestness and meaningfulness.

Once, late in 1877 when father was 26 and they were all sitting at the dinner table, Hubrecht sprang up and to everyone's surprise ordered a bottle of champagne. When he had poured for everyone present, he said, "I'd like to drink a toast to

---

[1] The Englishman William Herschel (1738–1822) was a Hanoverian emigre and itinerant musician turned amateur astronomer. Discovering the planet Uranus in 1781, Herschel became the most celebrated astronomer of the early nineteenth century and is know today particularly for his work in stellar motions and for his project 'Construction of the Heavens'.

[2] The English astronomer and brilliant mathematician A. S. Eddington (1882–1944), who was appointed Plumian Professor at Cambridge in 1913, made major contributions in statistical astronomy, astrophysics and stellar structure, and general relativity and cosmology, becoming one of the most influential theoretical astronomers of the twentieth century.

[3] Eddington, A. S.: 1923, 'Jacobus Cornelius Kapteyn, 1851–1922', *Proceedings of the Royal Society of London* **102**, xxxii (xxix-xxxv).

[4] W. Burck became a well-known Dutch botanist.

[5] A. A. W. Hubrecht (1853–1915) was professor of zoology and comparative anatomy at the University of Utrecht from 1882 to 1915.

Willem Kapteyn, who has been named professor of mathematics at Utrecht."[6] That was a surprise for Ko, who had not yet heard. Father rejoiced for his older brother, because a professorship was the most desirable position one could receive.

When the congratulations had settled down, Hubrecht ordered another bottle of champagne and after he had again filled glasses he declared, "And now I want to drink another toast to one who is present and is also going to be a professor. To Ko Kapteyn who has been named professor of astronomy at Groningen." This was an even greater surprise, not the least of which for father himself. As the son of the Secretary General of the Department of the Ministry of the Interior, Hubrecht sometimes knew things sooner than others. Both brothers were named by two royal decrees, '14 December 1877 No. 35, 36 Staatscourant 294'. It was a double celebration. After dinner they all went out singing and merrymaking through the streets of Leiden. While passing an old woman, who stood looking on and shaking her head at the loud young people, Hoek went to her and said, "And would you believe that one of them is a professor?" And he laughingly shoved father forward, who was certainly not among the quietest. "You don't say," said the woman as her hands fell to her side. Did professors look like that? She had imagined otherwise. And she stood there looking at them until their young voices were stilled by the darkness of the evening.

That same night old Mr. Kapteyn was holding a reading for the benefit of the community of Barneveld. Right in the middle of it, a telegram arrived announcing the appointment of his son Willem as professor at Utrecht. With great joy, he shared this news with his guests, and after everyone had congratulated him he went on with his reading. Shortly thereafter came a second telegram announcing the appointment of his son Ko. Great enthusiasm broke loose while the old father gleamed with joy and pride. The chairman took the floor and said, "Mr. Kapteyn, may I interrupt you and ask that you end your reading and tell us something of your life." And that he did. Being the great story teller that he was, it was a great improvisation of his own life.

My father was 26, an age at which most were either still finishing or had just finished their university work. While his studies had created great expectations, people were not surprised at this unusual appointment.

On 20 February 1878 father first came to Groningen. The situation, however, did not look very rosy for a young and enthusiastic professor of astronomy. Because the department chair had only been recently installed, father found that there was no observatory and very little chance of getting one. Imagine, a professor of astronomy with no observatory; that was unthinkable and totally unheard of in those days. Still, what was not could yet come to be, and with energy and unquenchable optimism father set about building an observatory. As a result, many letters were written, elaborate plans were laid, and numerous proposals were made. Father could get little done, however, except to locate a place outside of town for an

---

[6] Willem Kapteyn (1849–1927) became professor of mathematics at the University of Utrecht in 1878 where he remained throughout his professional life.

Fig. 3. J. C. Kapteyn and Elise Kalshoven, ca. 1878.

eventual observatory. Still, nothing further came of it, because he lacked support from within the government. The astronomers H.G. van de Sande Bakhuyzen[7] at Leiden and J. A. C. Oudemans[8] at Utrecht both gave the Minister misleading advise, and argued that a third observatory[9] in Holland would be one too many. The answer from the disappointed young astronomer was bitter: "Because of your advice, for an indefinite period of time Groningen shall be denied the proper instruments. In all of the Netherlands, the astronomers in Groningen are the only ones who do not have a proper work place for scientific research." To no avail, neither support nor money could be found, and after a last vain attempt in 1890 to get a photographic dome with a refractor from the government, father gave up. Because of these circumstances and his own genius, father found new directions to follow that resulted in the formation of the Astronomical Laboratory. But many battles and disappointments would still have to be endured before he would come that far.

During the summer of 1878, he asked Elise Kalshoven to marry him. Both knew that it was good, and being full of trust they looked forward to their future. The engagement lasted a year during which time father frequently traveled to Utrecht. Even in all their joy, the young couple were serious-minded. They talked about the big dreams that father promised, which were also seriously considered by Elise.

---

[7] H. G. van de Sande Bakhuyzen (1838–1923) was professor of astronomy and director of the observatory at the University of Leiden (1872–1908).

[8] J. A. C. Oudemans (1827–1906) was professor of astronomy at Utrecht (1875–1898).

[9] The other two observatories were at Leiden and Utrecht.

A completely new world of the intellect and of science had opened up for her, a world far and above the level that she had known as a youth.

The revolution brought about by Charles Darwin had made an enormous impact on their times.[10]   It was a time of rationalism, and young people had begun enthusiastically to reject the old norms and dogmas. They felt like gods in the kingdom of the intellect where old conventional traditions were discarded like the lifeless branches of a perpetually green living tree. In Holland, Multatuli's voice thundered, and swept along with it all that was young and full of fire and not yet fallen asleep under the safe and trusted mantle of tradition. Father revered Multatuli and enthusiastically chimed in, "Freedom, the highest right!"[11]

Into this world of new ideas he led his future wife. The first book he gave her to read was John W. Draper's *History of the Conflict Between Religion and Science*.[12] Although the problems were far too much for her to handle, this book had a great influence upon her and filled her with much wonder. They both felt lifted up with a feeling of earnest and courage, where truth makes one free from the stifled traditions that people had been in bondage. They bought the complete 'Household Edition' of the *Works* of Charles Dickens[13], which became a source of happiness throughout their entire lives. The best of Dickens, this brilliant story teller with a great heart, was enthroned in the living room where his spirit lived in their home.

In the summer of 1879 they were married. He chose well by marrying this happy girl who he brought into his home. And she found her own place beside this great and good man.

Groningen

Because of poor transportation, Groningen was considered a remote, provincial little town in Holland, and therefore not really part of Dutch civilization. The stuffy Groninger, behind in fashion and the times in general, was not held in high esteem. Groningers, who were stiff and wooden and put on airs, spoke and dressed differently than in more fashionable Utrecht. Consequently, my mother, the young

---

[10] Charles Darwin (1809–1882), who published his *On the Origin of Species by Means of Natural Selection* in 1859, was the naturalist who independently co-discovered along with Alfred Russell Wallace (1823–1913) the modern theory of evolution.

[11] Eduard Douwer Dekker (1820–1887; pseud. Multatuli) was a novelist and essayist and the most brilliant Dutch prose writer of the nineteenth century. His masterpiece is *Max Havelaar of de Koffij-veilingen der Nederlandsche Handelmaatschappij* (1860).

[12] The two most influential anti-religious, pro-science books in the Anglo-American world of the nineteenth century were written by John W. Draper (1811–1882), American immigrant son of an itinerant Methodist minister, who published *History of the Conflict Between Religion and Science* (1875), and by the American Andrew D. White (1832–1918) president and founder of Cornell University, who published *History of the Warfare of Science with Theology* (1896).

[13] The complete *Works* of Charles Dickens (1812–1870) is fully twenty volumes in length.

Mrs. Kapteyn, did not feel at home here. To her it seemed as though she had come to a completely new land. In contrast, the very last thing that anyone could say about my parents was that they put on airs. They rented their first house in the Winschoterkade, a neighborhood occupied by sailors and simple laborers. No professor had ever thought about choosing such a humble neighborhood. But my parents had little money, and using common sense they chose a roomy house at an inconsiderable height over smaller, better ones, but were very happy.

Their circle of professors consisted of much older, sedate people. Mother and father were so young to have been taken in by such a respectable circle. Mother would stamp her feet out of irritation and impatience over the measured formality and tediousness of the older people. Why must professors and their wives be so formal and put on airs, she thought. Rather than letting herself feel intimidated, however, she felt doubly happy in her youthfulness, rejoicing that they could still be as happy children and live life as they saw fit according to their wishes, without being bothered by the demands of rank and position. They were young and happy, expressing themselves sometimes as children that are suddenly made lord and master. It was all sun and happiness and quickly they made good friends that were drawn toward the natural, simple, and yet genuine young couple.

Elise Kalshoven was once a quiet young girl, but Elise Kapteyn developed into a spontaneous, vivacious woman. When her younger sister Marie came to stay with Elise (my mother) she was constantly surprised. After Marie returned home she told her mother: "You wouldn't know Elise anymore; she's completely changed." While her originality and vivacious spirit were dampened somewhat at home, in her own element these qualities could freely develop. My parents displayed a new style, which many in Groningen accepted with a smile, perhaps enviously, as they observed the young couple.

After a year their first child, my sister, was born. Together they had made a thorough study of the principles of child-rearing. This was exceptional at that time, but they were ahead of their peers in their ideas. Many times they were able to set the midwife straight who was so used to giving out his own brand of wisdom regarding such things. This was indeed no small thing because in those days midwives commanded with an iron hand and did not put up with any argument. Father also set to work on the doctor. These being the generally recognized authority in such things, father would only follow along whenever his common sense agreed with theirs.

When only a few months old, my sister developed an intestinal infection. Although the doctor prescribed several kinds of proteins, father had the courage to disregard the prescription. He believed that the weak stomach of a nearly starving child could not tolerate proteins. Although keeping it limited to a minimum, father used a diet of sugar-water. Still, as my parents kept a fearfully tense watch, eventually it produced results as my sister regained her strength and health. What a responsibility they took upon themselves! As a result, they continued to trust their common sense. They weighed my sister regularly, which was not normally done, and they read books about feeding and child raising.

Both Allebé and Rousseau were thoroughly studied in order to gain a better understanding of child up-bringing. Dreams were dreamed, such as those dreamed by every couple at the cradle of their first child. In contrast to the foolish etiquette of the time, mother was the first in her circle of friends who pushed her own baby carriage rather than have a nurse-maid do it. And if they went out together, the young professor pushed the carriage himself, despite the laughs of the kids in the street, who were less tame then they are now. But the criticism of the street urchins did not harm him anymore than did the admiration of his colleagues. On the contrary, mother found it a pleasant challenge to defy convention, because she knew that her independence from traditions made her happier than others.

When my sister was a year old and when there was still no prospect for a second child, my parents judged it unwise that she should grow up alone. Being full of pedagogical principles and feeling that my sister would be happier in life if she had a playmate, they decided to find one and take care of it as well. Nearby a neighbor had gone bankrupt and had left for America. His poor wife, who had been left behind, found herself miserable with her child and complained of her distress. Father immediately stepped forward and offered to take the child in as his own in order to free this woman from her worries. Mother, who was less idealistic and more practical, was not very happy about taking a strange child into her family. As it was, she had her hands full with her fine, healthy little daughter. And besides they had to live very frugally. But the urgent assurances of my father about the pedagogical necessity finally won its case. The child came to live with them and, as my mother feared, the child became a big pain, because it had been dirtily and badly raised. Consequently the child was surprised at everything it saw happen in her new home. Fortunately, the child stayed only three months, because the mother came back to take her to America. Apparently, the child's parents had planned to arrange for someone temporarily to care for the child. My parents breathed great relief that this affliction would not be theirs for their entire lives.

Father also believed that children did not have a natural affection for dolls, but that this was urged on the child by its parents. Because of this, he would not allow his child to have one. My poor sister would have to go through life doll-less. Once, however, he saw my sister with an ugly, dark Japanese doll that was placed on a stick to be moved up and down and that was cradled in her arms as she rocked it like a mother does her child. Although he said nothing, he came back that afternoon from town with a package in which my delighted sister found a beautiful doll. His theory appeared to pale somewhat as the shining rays of life taught him otherwise. As a result he let his daughter go her own way. In the larger realm of life, particularly later during his scientific discoveries, this seemed a useful attribute. For example, sometimes a beloved, hard-won theory must be abandoned and discounted or later honorably amended. How many set backs does a true servant of science know of which the world knows nothing except only the results and the difficult ways in which they were obtained.

While in the house on the Winschoterkade where they lived six years, they had a son, my older brother, and another daughter, myself, born to them. Father was a natural educator just as his father had been. The children learned effortlessly at meal times and on walks; everything that he told them was interesting and sparkled with life. He placed importance on everything that concerned them and never ignored them with easy answers. In each of us he saw a developing adult, each having their own nobility that one day must be respected and well studied.

In educational matters father followed the principle of causality: Children must learn to understand and live with the consequences of their actions. Life punished the offender, not the father, and in this manner my father prepared his children for the life that awaited them. His seriousness and dedication, his patience and love, his healthy understanding and practical outlook all combined to form a rock solid faith in him. He was a never-failing or disappointing security for us, but he was also a jolly comrade and a tender friend.

Our mother stood gallantly by his side, softening the sharp corners of his character (which his lonely youth had given him) by her warmth, spontaneity, and unselfish love. She was always busy, and worried trying to make ends meet, which indeed was difficult in those first few lean years. She was jolly and happily not logical or exact; a laugh and a joke at her expense was something she endured well.

Even though sometimes they found one another difficult to understand, her playful originality charmed his intellectually critical nature. Although their dispositions and backgrounds were also quite different, slowly they grew more and more like one another because they both had something to give and both had the desire to learn.

In order to see his scientific plans brought to fruition, however, father suffered under almost impossible circumstances. Therefore, everywhere he sought for work with his hands and head. In those early years, he occasionally collaborated with his mathematician brother Willem whose ideas in higher mathematics eventually saw the light of day. Father also spent his vacations at the observatory in Leiden where he relied on the hospitality of its director van de Sande Bakhuyzen in order to use the instruments for research. In 1883 he published his first astronomical paper.[14] His method later proved useful by improving fundamental observations. Later he used the meridian circle for research in stellar parallax.[15]

Meteorological studies also kept him busy, traveling to Worms and Paris in order to do research on the growth of trees in relation to weather conditions. He inquired of the Dutch government if they would ask the French government for the locations of trees at least two hundred years old near Paris. There was the meteorological station that had kept the longest records of meteorology, especially regarding rainfall. His research regarding growth of trees kept him busy for awhile,

---

[14] Kapteyn, J. C.: 1883, 'Über eine Methode, die Polhöhe möglichst frei von systematischen Fehlen zu bestimmen', *Copernicus* **3**, 147–82.

[15] Kapteyn, J. C.: 1891, 'Bestimmung von Parallaxen durch Registrier-beobachtungen am Meridiankreise', *Annalen Leiden* **7** (3), 117–244.

but did not lead to any theories. Only much later did he publish the results.[16] None of this wholly pleased him; this was really just child's play. He wanted to do far greater work, which he knew he was now ready to tackle. Suddenly, however, came some opportunities that would give his life an entirely new direction.

---

[16] Kapteyn, J. C.: 1914, 'Tree Growth and Meteorological Factors', *Rec. Trav. Bot. Neerl.* **11**, 70–3.

CHAPTER 3

EARLY PROFESSIONAL YEARS

The 'Cape Photographic Durchmusterung'

During Christmas vacation 1885 father was reading an article written by David Gill[1], Her Majesty's astronomer at the Cape of Good Hope, entitled 'To Lead' in an astronomical journal about a big project that Gill was hoping to undertake. The project entailed making a catalog of all the stars of the Southern Hemisphere up to the tenth magnitude using photographic techniques. There was a tremendous amount of work that still had not been done. For the northern hemisphere the *Bonner Durchmusterung*, completed in 1859 under the leadership of Friedrich Argelander[2] of the Bonn Observatory, already existed. The latter was a catalog of 324 000 stars up to magnitude 9.5 each of which had been recorded visually between the years 1852 and its completion in 1859. Although the positions were provided with reasonable accuracy, the *Bonner* catalog was of the greatest importance to astronomers. It was virtually the only source for research for 'the construction of the heavens',[3] which entailed studies of the density[4] and luminosity[5] distribution of the stars, as well as the basis of a great number of newer research problems. More than any other, this catalog contributed to an understanding of the movement of the stars. As a result of the *Bonner* catalog, Argelander brought to light many of the mistakes of other stellar catalogs. Above all, astronomers found this work a faithful and reasonably complete representation of the Northern Hemisphere as understood in 1860.

The idea occurred to my father, "Here is a chance for me!" He immediately wrote Gill and offered his help: "If you will confide to me one or two negatives,

---

[1] David Gill (1843–1914) was Her Majesty's astronomer at the Cape of Good Hope from 1879 to 1906; see pages 31–34.

[2] The German astronomer F. W. A. Argelander (1799–1875) spent his most important years as professor at the University of Bonn from 1836 to his retirement and while he was director of the Observatory he produced the *Bonner Durchmusterung* between 1852 and 1859. His main astronomical contributions consisted in providing complete data on the positions and magnitudes of the northern hemisphere stars down to magnitude 9.5. Without this work the statistical cosmology of Kapteyn would not have been possible. Later F. Schönfeld also at Bonn extended the *Bonner Durchmusterung* to −23° declination in the southern hemisphere.

[3] 'The Construction of the Heavens' is the name William Herschel gave to his project for understanding the nature of the Milky Way system. For details of his project see Hoskin, Michael: 1963, *William Hoskin and the Construction of the Heavens*, W.W. Norton & Co., Inc., New York.

[4] The 'density' relationship provides values for the absolute numbers of stars per unit volume of space as a function is stellar distance.

[5] The 'luminosity' relationship provides values for the spread of stellar brightnesses by magnitude class and stellar spectra.

I will try my hand at them and, if the result proves as I expect, I would gladly devote some years of my life to this work, which would unburden you a little, as I hope, and by which I would gain the honor of associating my name with one of the grandest undertakings of our time."[6] Gill answered: "It is not easy to tell you what I feel at receiving such a proposal. I recognize in it the true brotherhood of science and in you a true brother."[7] And so these two men found each other, who as true colleagues would work and struggle side by side. They were like two titans who would assault the heavens together, but with larger goals than the titans of ancient mythology.

The astronomer van de Sande Bakhuyzen and his brother Willem at Leiden, however, were not very enthusiastic. Naturally the end result would be of tremendous worth, but the slave work, indeed, was almost unbearable. Father wrote to Gill: "However, I think my enthusiasm for the matter will be equal to (say) six or seven years of work."[8] Indeed, his part of the work would demand a great deal of enthusiasm, because it entailed measurements of star images on photographic negatives. The reduction of the star-plates, which would definitely need more than twenty times the work spent on the photography, was now a possibility for him to undertake. Without an observatory, however, such work was expensive and tedious. For the development of knowledge it was essential to perform important work from which he hoped to earn his reputation and thus honor the University. In the meantime it was not feasible to achieve work of such proportion without help, even with the help of student astronomers.

From the government he obtained a small grant of 500 guilders for a period of seven years for which he could get the needed assistance to execute most of the work involving the photographic machines and for doing the written work and mathematical reductions. The grant was awarded him for six years. He still needed more resources, however, for he did not yet have a 'pied a terre' and no measuring instrument. Then his friend Huizinga[9] offered him two workrooms in the latter's laboratory, which father gratefully accepted. When he got his own laboratory a decade later, turning to Huizinga, father expressed how he felt: "When I came to you I had known hopeless, the feeling of a useless human being. When I left you, free of that feeling, I thanked you for that. Who can ever calmly lay down his head unless one has the feeling that he has not lived completely uselessly?"

Because so little money existed, the building of an instrument required much ingenuity. Thus he inventoried the existing instruments, which were mostly old and fairly useless, that were at his disposal. From this he succeeded in creating

---

[6] J. C. Kapteyn to D. Gill, 16 December 1885, reprinted in *Cape Photographic Durchmusterung*, i (1896), p. xiii of the *Annals of the Cape Observatory*, iii (1896) (original letter lost).

[7] D. Gill to J. C. Kapteyn, 9 January 1886 and 22 January 1886 (K.A.L.) in which Gill gladly accepts Kapteyn's generous offer for assistance.

[8] J. C. Kapteyn to D. Gill, 16 December 1885, reprinted in *Cape Photographic Durchmusterung*, i (1896), p. xiii of the *Annals of the Cape Observatory*, iii (1896) (original letter lost).

[9] Huizinga was professor of psychology at the University of Groningen.

an instrument composed of different pieces based on a whole new principle. The instrument, which was built under father's guidance by a mechanic from Groningen, satisfied father completely. It worked to such a degree that later the Commission of the *Carte du ciel* at Paris contracted to have one made for them for their own research.[10]

Still father needed an assistant. But because he could not pay a high wage, he decided to look for an unskilled worker. But where to look? Intelligent and talented men who wanted to work hard for little money were difficult to find. But he had an idea: he went to the Director of the Ambachtschool and asked him if he could recommend a good student, suited for this type of work. The Director knew of such a student and gave father the name of a nineteen year-old who had the necessary qualities. Although the boy T.W. De Vries came to him as a temporary assistant, he immediately undertook the measurements with the instrument and performed them excellently. As an excellent observer, he seemed to have a special talent for accurate measurement work; even the calculations he completed with diligence and accuracy so that he quickly became indispensable. Yet father knew very well that De Vries, with his skills, deserved a better position and a better salary than he could ever expect from such a modest beginning. Therefore, in 1901 father introduced De Vries to his colleague Albertus Nijland[11] in Utrecht, at whose observatory a better salary was forthcoming. After he had mentioned De Vries with great praise, he noted: "Heaven knows that with great regret I shall see this man leave, but I don't feel free to keep a career from him. Such a chance might not come again so quickly."

Fortunately, because of his splendid resume provided by the government, in 1911 father offered De Vries the job of amanuensis in his laboratory. For many years, De Vries remained father's right hand until the latter's retirement, when De Vries continued to work another six years for my father's successor, Pieter van Rhijn[12], in order to be able to profit from his well-earned vacation. The laboratory owes much of its success to De Vries.

My father and De Vries worked together in the small rooms of the psychology laboratory with the newly designed instrument. The best accuracy was achieved when each star was measured at least twice, the second time by father himself. The work progressed more quickly than expected. As Gill wrote Kapteyn in 1886: "You

---

[10] Beginning in 1889, the *Carte du ciel* enterprise resulted in the cooperation of an international group of observers for an atlas of the entire sky down to the fourteenth magnitude and a catalogue of exactly measured star places down to the twelfth magnitude. Because the plates were only 2° square, the number of plates needed was so large that the project was only recently finished. For considerable material dealing with this project, see Debarbat, S., Eddy, J. A., Eichhorn, H. K. and Upgren, A. R. (eds.): 1988, *Mapping the Sky: Past Heritage and Future Directions*, Reidel, Dordrecht.

[11] A. A. Nijland (1868–1936) was professor of astronomy and observer at the Utrecht Observatory. He wrote a mathematics dissertation under the direction of Kapteyn's brother, Willem Kapteyn.

[12] Pieter J. van Rhijn (1886–1960), who was one of Kapteyn's eight doctoral students, succeeded his mentor as director of the Astronomical Laboratory of Groningen following Kapteyn's retirement becoming professor of astronomy at the University of Groningen.

have made an excellent beginning to the work, the results far exceeding in accuracy what I thought you could attain with your provisional apparatus." Indeed the accuracy was noticeably greater than was achieved in the *Bonner Durchmusterung*.

The small grant from the government was really not sufficient, however. With despair father wrote Gill, who then wanted to get money in England. Even though my father was thankful, he could not accept it for the honor of his University and his country. So he decided to find another way to obtain the needed funds. He contacted a society in Holland that financially supported research, and, as a result, he obtained from Bataafsch Genootschap 150 guilders for several years. But that still was not enough. He then tried to get a grant from the government of 500 guilders for six years that was later changed to 1000 guilders for 3 years. "The fourth year I will be sadly at the end of my resources," he wrote to van de Sande Bakhuyzen, "but until then there still might be some other way." And his optimism, a wonderful quality in his life, did not let him down, because after those three years he had done such fantastic work that his name became known world-wide so that he could pursue whatever research he wished.

Still in 1887 it looked bleak for Gill and my father, who had quickly become good friends. Gill who always hit the nail on the head wrote Kapteyn: "Surely if in this wicked world there ever were two men who should be good friends together, you and I should be so."

For the two of them, their correspondence was the source of much happiness with each letter received enthusiastically. When in 1887 Gill was to come to Groningen for the first time, our family hovered in suspense. Each prepared himself. My mother, who spoke excellent English, briefed her children – aged 5 and 7 years – so that they would have a better understanding that would be in debt at their meeting. When Gill arrived everyone liked him. He had the gift to win all our hearts through his honesty and enthusiasm. During the first evening Gill and father sat in front of the fireplace where Gill told his life's story from which his simplicity made a deep impression. The children learned a beautiful lesson in life from him that not many understand: Give trust and you shall win trust, and as a result life becomes richer.

Mother doted on the twosome, who went to the laboratory the next day. The Groningers stared at Gill, who talked and gesticulated so much, while father, somewhat small and petite, was quietly happy next to him. Their first visit was a splendid success. Even we children were impressed by this big-game maker, who had won our hearts in an incomprehensible language. From that time forward we called him 'Uncle Gill'.

Many ideas and issues were discussed since they had much to struggle with, particularly the troublesome finances that did not look promising for their work. William Christie,[13] the Royal Astronomer at the Greenwich Observatory, was also

---

[13] While W. H. M. Christie (1845–1922) was Royal Astronomer from 1881 to 1922, the observatory at Greenwich prospered more than it had done since its founding in 1675. Christie is known mostly for advancing the administration of astronomy, however, than in advancing its theoretical foundations.

deeply concerned that Gill would overshadow him. Because support for their work depended on Christie, the latter did his best to make the cataloging work nearly impossible. Articles about their work appeared in scientific magazines. Because there were so few stars on the plates that were sent to the Royal Society as proof of their success – many fewer stars than even on the plates of the Milky Way of Paul and Prosper Henry[14] of Paris – there was opposition to their *Cape Photographic Durchmusterung*.

If it was not so sad, one could laugh about it. For this reason, however, astronomers tended to reject their work. Writing to the American astronomer, Simon Newcomb, Gill noted: "I told them that I had heard of babies crying for the moon, but I had never dreamt of anything so funny as a row of Fellows of the Royal Society insisting on having more 9.5 magnitude stars in the heavens, else they would stop supplies."

Articles also appeared in the astronomical magazine *Observatory* in order to urge Gill to leave his post and to plant seeds of distrust between Gill and my father. On this, Gill commented to father: "I wish that the stupid fellows who are writing silly abuse of me in the *Observatory* would take a lesson from you and instead of doing that would really work at improving existing methods of measurement or inventing new or better ones." But the two stood firm and nothing could hinder their working together.

Gill proposed to pay expenses out of his own pocket, and his wife offered to sell her riding gear. "We will somehow manage it together, even if we have to give up our tobacco – which God forbid!" Friedrich Auwers[15], the astronomer from Berlin and a friend of Gill, also offered help. As a result, a Humboldt grant[16] from the Berlin Academy of Sciences took care of financial lacking, which was much needed. But Gill and his wife, who indeed made all kinds of cutbacks, put much aside from their work that everything worked out alright.

In June 1892, father wrote Gill: "Finished! The job of measuring the plates done, done at last. The work has been to me a source of no end of good things, but still its being done at last is one among the best ... the number of observations we got, must be upwards of a million – and the truth is that I find my patience nearly exhausted." Gill's answer to father the following month was a short letter that could only just make it in the mail: "This is only a jubilant shout – a hurrah – a God bless you, my boy, – and long may you go on and prosper!"

In that same year father was named a Foreign Member by the Royal Astronom-

---

[14] The brothers Paul Henry (1848–1905) and Prosper Henry (1849–1903), who worked together throughout very productive lives as technicians, built unsurpassed photographic telescopes one of which in 1887 became the prototype for the international project of the *Carte du ciel*.

[15] A. J. G. F. von Auwers (1838–1915) was astronomer of the Berlin Academy (1866–1915). From 1865 to 1883 he completed a new reduction of James Bradley's Greenwich observations which formed a foundation for research of star positions and proper motions. He was a key participant in the founding of the Astrophysical Observatory at Potsdam.

[16] The Humboldt funds were named after the naturalist and explorer Alexander von Humboldt (1769–1859).

ical Society, a high honor that gave him much happiness and satisfaction, and that stimulated him to complete the demanding work of finishing the calculations in the star catalog that still awaited him. It was difficult work, to which his family could testify. Otherwise our sociable father and companion, who always told jokes and played games, who made statements and asked questions, and who told exciting stories especially during meals, now had no time. Meals were quickly eaten so that he could hastily get back to work.

He had neither zest nor freedom. For a long time he got up at 4 o'clock in the morning so that he could be at the laboratory at 5 o'clock. Then he worked until 9 o'clock and rested until noon so that he could again begin with renewed strength. This regime suited him well, though. He acquired enormous strength through which he could accomplish much.

Mother lived with him completely. She also got up at 4 o'clock in the morning, organized her housework to suit his work, thought of new and delicious meals with a resourceful mind, and removed all difficulties therefore helping to increase his work and efficiency. She did nothing for herself, because she knew that the tide wasn't subject to change. At the end of all this work, the rest and coziness of earlier times would again return. She was a help to her husband in the true sense of the word, without question and unconscious of her own efforts.

But the bow was too tightly strung. Father became agitated and would then fall apart back at home. It was really no wonder that his humor suffered under these enormous demands that were spent on his challenges. But we children could not understand the deeper reasons, and we thus felt unhappy and hurt. Mother discussed it with father but they blamed themselves. She seldom complained and asked nothing from him, but it was because of this that the significance of the complaint made a deep impression on him. He called us children to him and told us that he had done wrong. The happiness in the home came above all else in the world; consequently, he promised to do his best and behave himself. He spoke long and seriously, he humbled himself before his children as only an adult can do. At that moment he not only won back what had been lost, but he awakened within us a deeper understanding and admiration. Because there aren't words for these emotions, we didn't say much, but we never forgot this moment.

Physically the tension was also too much, because he had acquired a nervous stomach and had achy eyes. He couldn't even think of resting. Sometimes he lay stretched out on the floor with books spread out in front of him because it was the only way he was comfortable. As a result, father became thin and slept badly. Although he was consumed by his work, the writing of an endless number of letters became his main outlet. For instance, to the astronomer Anders Donner[17], whom father had met during the 1890 *Carte du ciel* congress in Paris and with whom he became friends, he wrote, "I am being consumed by the work of the Photographic Durchmusterung, and that makes me put aside everything else."

---

[17] The Finish astronomer A. Donner (1854–1938) was professor of astronomy and director of the observatory at Helsingfors (1883–1915).

The Durchmusterung had indeed consumed him. In 1899 it lay in front of him in three quarter parts. The work had taken twelve years, double what he had thought in the beginning, but still it was of utmost importance to astronomy. As Gill wrote my father on 16 April 1899: "A thousand heartiest congratulations on the completion of the Durchmusterung catalogue. What a load off your weary shoulders! How splendidly you have redeemed the promise you made me in 1885 and how thoroughly you have done your great work! It will ever remain a standing memorial of your devotion to science, your earnestness of purpose and your wonderful working capacity...." The work was reward in itself, because it had shown my father what he could undertake, and it had earned him the celebrity and admiration of the whole astronomical world.

As Newcomb commented: "This work of Kapteyn offers a remarkable example of the spirit which animates the born investigator of the heavens. Although the work was officially that of the British government, the years of toil devoted to it were expended without other compensation than the consciousness of making a noble contribution to knowledge, and the appreciation of his fellow astronomers of this and future generations." Last but not least the work had given my father a friend for life, whose rich experience and wisdom had a large influence on him, and whose honesty and warmth had won his heart. Writing to my father, Gill noted: "I congratulate myself that the material furnished to you – however many its imperfections – has enabled you to establish for yourself a reputation and position amongst the astronomers of your time such as few men of your age enjoy. Above all I rejoice in the true friend I have found in you – may that friendship ever grow with our years!" Although the work did not bring financial advantages, the harvest was so rich that he felt himself a pardoned human being. The feeling of uselessness was forever gone; he had not lived in vain.

David Gill

In the beginning of his career, father was very lucky to have found Gill who had such a great influence on his entire life and work. A noble man, a great astronomer, a good husband, and a true friend. "Of such is the salt of the earth!," exclaimed General Charles Gordon[18], the famous hero of Khartoum, when he first met Gill at the Cape. And so thought everyone who met him.

The son of an Aberdeen clock maker, David Gill was destined to follow his father in the family business. But astronomy captured him, and he who has knowledge does not easily let go. And so it was with him. Consequently, he disposed of the business with all its security and followed his love of knowledge. In 1872, he became director of a small observatory in Dun Echt, Scotland. He organized several astronomical expeditions to the Mauritius and Ascension islands,

---

[18] Charles George Gordon (1833–1885) was the British general who became a national hero for his adventures in China and his defense of Khartoum (1884–85) against Sudanese rebels during which he lost his life.

Fig. 4.   David Gill, 1884.

and, through his skills as a practical astronomer, he made himself a name. In 1879, when the position of Her Royal Majesty's astronomer became vacant, Gill applied for and was named to the position. He remained twenty-seven years in this position during which he raised the observatory at the Cape of Good Hope to great heights. In 1906, largely because of the strain due to his active professional life, he resigned settling in London where he became astronomy's middleman. Shortly before, on 24 May 1900, he had been knighted Sir David Gill, because England knows how to honor its great men and to distinguish them from others.

As an honorary member of the Amsterdam Academy of Science, many benefitted from Gill's decisions. Nearly everybody sought his advice because of his
extensive experience and because his clear insight was unequaled. From around
the world scientists visited him at his cozy flat in London at 34 de Vere Gardens,
Kensington. With enthusiasm and dedication he would become lost in each issue
knowing exactly how to help and give advice. A truer friend and impartial admirer
did not exist. He possessed a beaming enthusiasm and youthful delight in what
was large and good, a deep religious feeling, and an irresistible sense of humor.

Many anecdotes are known about Gill. Polite, intellectually modest, and
touched with the gift of genius, Gill still had a proper sense of self-respect. This
became apparent at the 1890 Congress of Astronomy in Paris. Following the voting
for president of the Commission of the *Carte du ciel*, it was unanimously decided
for Gill. "Yes," he said without reluctance, "I found myself the most qualified
person," which without doubt he seemed to be. Greatness belongs to him who can
do something like this without scandal.

Gill was one of those rare people who could do anything and everything without
ever hurting anyone else. He was liked all the more because of this wonderful
quality. When he met George Ellery Hale[19], the handsome director of the Mount
Wilson Solar Observatory for the first time, Gill asked him, "What are you going
to do with that five foot reflector?" After Hale laid out his plans, Gill responded:
"All wrong! You should do nothing but radial velocity work. Now go ahead and
defend yourself." "The twinkle in his eye," wrote Hale about him, "overcame any
fear of aggressive intent, and the cordial interest he showed was characteristic of
the man."

His married life was ideal. Lady Gill was also worthy of him. She was of
high standing and an intelligent woman, who both shared his sorrow and gloried
in his triumphs. During their engagement she quietly studied astronomy. When
her interest became known to him, thinking to see him surprised, Gill exclaimed:
"All exploded notions!" Intelligent woman that she was, she relinquished all her
attempts and came to believe that a life time of study was needed to stay current
and up-to-date. Later, when someone once asked Gill, "I suppose your wife knows
all about astronomy?", he answered from the bottom of his heart: "Not a word,
thank God!"

Their relationship was probably best expressed in a letter to Hale, thanking him
for an invitation to visit Mount Wilson: "The sad state of my wife's health which
Kapteyn will tell you all about, is such that I dare not go to the Solar Congress. It
is a very bitter disappointment to me, but there are things dearer to a man than any
congress, any gratification of friendship or the desire to see and know."

---

[19] George Ellery Hale (1868–1938) was the founder and first director of the Mount Wilson Observatory, which, in addition to becoming an international laboratory for solar research, Hale's own
specialty, it was dominated by the first operational world-class 60″ and 100″ reflecting telescopes.

After Gill's death, Lady Gill wrote to her husband's biographer George Forbes[20]: "Twenty six North Silver Street was a comfortable, but rather ugly little house, and the furniture which I thought beautiful, and David did not think about at all, atrocious. But to us both a very heaven of happiness lay between its four walls, as it always did between every four walls which held us two to the end of his life."

My father himself wrote the closing words of this beautiful life: "In many a human heart his image will last as long as life itself."

The Astronomical Laboratory at Groningen

Although grand and useful, the cataloging work could not free my father's mind. Still, he even had a bigger goal, namely the 'Construction of the Heavens'. All of his research, from the earliest to his last, was consumed by this idea. As vastly different as his research and publications may appear, they were all pieces of a much larger whole: to wit, they were all needed eventually to understand the architecture of the universe. Whatever details he may have been devoted to for the moment, he never relinquished his quest for the solution of this great problem. Ultimately, this was the key to his success. Indeed, many saw in my father his ability simultaneously to focus on the big picture and to devote careful attention to the smallest of detail. At the same time my father possessed 'the faculty to neglect the insignificant'. The combination of these three factors is hard to find. It is definitely an element of success in the sciences.

At the time that he was working on the catalogue with Gill, father was already forming plans for future research. In the meantime, however, much was happening. Already at the Paris congresses of astronomy in 1889 and 1890, men had become aware of the excellent work being undertaken by my father. As a result, he was chosen to serve on the permanent committee of the *Carte du ciel*, a huge international undertaking designed to determine the positions of nearly 1 500 000 stars spread over the entire sky. My father was asked to design the parallax instrument that he would use.

Both Gill and my father had considerable influence at the 1890 Congress, which displeased the Royal Astronomer William Christie. Consequently, the latter did his best to oppose Gill. For example, Christie argued that while Gill's proposal to have a central office of the *Carte du ciel* erected would certainly be very beneficial, it would be seen as menacing to the British participants. As a result, Gill had to withdraw his proposal. Despite its grandiose beginnings, however, this is certainly one reason why this enterprise remains to this day languishing and will continue to do so for years. Following Gill's death, father wrote: "How different would be the outlook now, if he could have carried through his plan for a central bureau, perhaps the only important measure which he failed to see brought about." After Gill's attempt failed, my father noted: "They said that Gill and Kapteyn would run

---

[20] A physicist by training, George Forbes (1849–1936) wrote some popular science materials including *David Gill, Man and Astronomer* (1916).

Fig. 5.   Jacobus Cornelius Kapteyn, 1914. (Courtesy of Yerkes Observatory.)

the whole show. There was not the slightest foundation for this beyond a horrible jealousy.... To some men the working for science and truth for their own sake is a thing they cannot understand." Even in astronomy, the science of unmeasurable greatness, one can see that pettiness is possible.

The two friends worked hard for their mutual interests. They infrequently

Fig. 6.   Astronomical Laboratory at Groningen, ca. 1910.

noticed obstacles placed in their way, and defended their ideas proudly. Consequently, they obtained excellent results and their work was well on it's way to
completion. Shortly after the Congress, father received the Legion D'Honneur, his
first distinction that brought considerable happiness to our family. We had every
right to be proud, because it was the first official acknowledgement of father's
abilities, soon to be followed by many others.
    My father now judged that it was again time to make a proposal to the govern-
ment. In May 1892 he wrote to the curators:

> The Groningen astronomy group possesses: (1) the measuring instrument with
> which the nearly completed photographic survey of the Southern Heaven was
> measured, and (2) the nearly finished measuring instrument which the Comite
> Permanent de la Carte du Ciel told me I could use. It was built after an earlier
> one but executed with the most attainable perfection. Furthermore, Professor
> Huizinga has offered to allow me to set-up the instruments in his laboratory,
> which is in every detail suitable for our needs. If we were to be given a
> Repsold measuring instrument for about 2400 guilders, then this astronomical
> institution should be in possession of a unique collection of photographic
> measuring instruments ... and, thus, here in Groningen, there shall be, better
> than anywhere else, even abroad, the possibility for studying the different
> methods and measurements of photographic star recordings.

His request for the instrument was approved and immediately ordered. It was
not immediately set up in the physics laboratory, though, because father had been

promised his own building, the temporary empty home of the Commissioner of the Queen.

On 16 January 1896 the laboratory was opened. Finally astronomy at Groningen had her own temple where the holy fire could burn, and where the young could be initiated into the mysteries of knowledge. My father wanted to give a worthy dedication to his astronomical laboratory. This would have been an incomprehensible affair to the large audience, because his astronomical work area did not have a telescope and a dome. Therefore in his inaugural address he explained:

> People count the photographical observatories by the dozen. At each observatory much more is produced than can be analyzed, because the work force available for measurements and all the other work, while adequate for the photographical recordings, is insufficient for data reduction. ... The curse of most observatories, which are most diligent in their work, is the continuously menacing accumulation of raw data. A number of the most progressive among them still commit to their archives data which they couldn't reduce, or which was insufficiently reduced, and thus it remains unfruitful for science. Is it any wonder that the observatories feel pressured because of this, and is it not time that people reexamine the imbalance between the collection of observed data and its use, between the work of the photographer and of the astronomer, in order to mend it at least a little bit? To devote themselves to what Darwin wrote so characteristically in his letters: "The grinding of huge masses of facts into law." I think that all will answer that affirmatively. With that the question is answered: an institution like this is justified to the world.

> The second question is 'Where does the material come from?' But the answer to this one is easy after the first one: because of an over production of observatories. Although many will be inclined to distance themselves from the material they have collected, to see it analyzed there can be little doubt. Useful work, therefore, is not lacking.

My father was thankful for this first step. But he could not recken himself among the "satisfied professors"; there remained much to be hoped for. The laboratory was only for his use temporarily, therefore no permanent pillars were permitted that were necessary for setting up various instruments. As well as the lack of a good technical library, many observational instruments were also lacking. "Without a library, an astronomical laboratory is not possible," he explained, "especially not as this one which, as stipulated in the charter, does theoretical research based exclusively on material collected by others. The publications of the more important observatories belong here, as well as other literature which should be well represented." Still, the first step toward establishment was taken when my father's laboratory was elevated to official standing with its own building and annual funding. With the motto, "Le mieux est l'ennemi du bien," father accepted his astronomical laboratory.

From everywhere be received appreciation. J. A. C. Oudemans, the professor from Utrecht who had failed to support father in order to help him realize the Groningen Observatory, later wrote:

> It is, after all, the question. Wasn't it really better without an observatory. When I was still an observer, that must have been in 1853 or 1854, I had to write to George Biddell Airy.[21] I told him that the plan was to build a new observatory in Leiden, and what was his answer? "Your message, that a new observatory will be built, I am not particularly enthusiastic. There are plenty of observatories. But an office of mathematics – that is needed!" The bureaus of mathematics in Berlin and Groningen have proven that judgement was not entirely uncalled for. But for good progress is the drive and perseverance that you possess necessary.

Now it is well-known that father's astronomical laboratory needs no more justification. It has brought forth big things and has a place of honor in the scientific world. In 1922, shortly before my father's death, the French astronomer Jules Baillaud[22] noted in his opening remarks to the International Astronomical Union in Rome: "The three things that have revolutionized the appeal of astronomy in the last half century are photography, telescopes, and [Kapteyn's] Laboratory in Groningen."

It may seem peculiar that until now an astronomical laboratory, not withstanding its great need, has remained a rarity. Only Professor Anton Pannekoek[23] of Amsterdam has constructed such an institute – in 1921, thus a very recent date. Its rarity, however, has deeper origins. The cooperation and the distribution of the work demands great difficulties, that require the utmost cooperation, unselfishness, and helpfulness of those who work together. For its material, a laboratory is completely dependent on observatories; a continuous understanding and good will is imperative.

But my father was the man for this challenge. Through his noble character (which knew no pettiness), his love for science, and his simple heartiness, he left to others the honor of their work. This environment became an ideal working arrangement from which he made the people he worked with his friends. Some called him diplomatic, but diplomacy was far from him. Diplomacy is using tact to reach certain goals, to play out certain factors against one another, and to make use of the vulnerability of others. For my father it was his unfeigned interest in the work of others, his winning selflessness, and his lovable modesty, all of which combined together for the benefit of those who worked unknowingly of his virtues. In every situation my father's keen mind perceived the combination of people and

---

[21] G. B. Airy (1801–1892), who from 1835 until 1881 was England's Astronomer Royal, had a major influence on the development of British astronomy beginning with his appointment at the University of Cambridge as the Plumian professor in 1828.

[22] J. Baillaud (1876–1960) was director of the Toulouse Observatory and head of the *Carte du ciel* at the Paris Observatory (1922–1947).

[23] Anton Pannekoek (1873–1960) was a founder of modern astrophysics in Holland, contributed to statistical astronomy, Kapteyn's specialty, but today is best known as having written his well-known *A History of Astronomy*, George Allen & Unwin Ltd., London, 1961.

strengths needed for success. No effort was ever too much for him and no obstacles could hinder him. In short, no one could resist his genius and personal charm. This is something completely different, and certainly much higher and nobler than what diplomacy could ever achieve.

Still, he sometimes felt the difficulties that came with the dependency of his laboratory on the work of others. So, in 1914, father wrote Sir Frank Dyson[24], Director of the Greenwich Observatory:

> I sometimes cannot help thinking that you must take me for the most trouble-some man alive. The man who has always something to ask and has never something to offer in return. I hope you will consider that it is not all my fault. It is the one point that makes such an institution as an astronomical laboratory objectionable, which must live on the materials furnished by other institutions. For several of the observatories which have helped us, the laboratory has been able to render small contra-services and we hope that some day will come, that we can do this for Greenwich....

The work of the Groningen Laboratory was twofold. First, the theoretical research and calculations needed for the 'construction of the heavens': The mean-parallaxes, the luminosity law of the stars, and the number of stars in each magnitude class. Second, the practical researches: The measurement of photographic plates in order to obtain data needed for the general research about the structure of the stellar system. All of this work has appeared in the many issues of the publications of the Astronomical Laboratory at Groningen.[25]

In 1889 my father developed a technique to determine the parallaxes of the faint stars as a group by recording them photographically. These recordings were extremely laborious. Using the same plate, each stellar group had to be photographed three times, each half-a-year apart. Thus each plate had to be exposed and then put away three times, before it could finally be developed. This was the same general technique father used for a much larger expansion of this plan that he introduced in 1904.[26] The movements of the stars were to be determined by looking at the plates from which two recordings were to be done at a 5–10 year interval.

When father discussed this plan with Donner in Paris, he was at once prepared to take upon himself the entire effort needed for this huge scientific goal in order to get the most precise measurements and the most complete results hitherto achievable. Father himself did the measurements and the analysis. By working together, Donner, who visited my father a number of times in Groningen, and father developed a warm friendship. Because both were poor correspondents, however, their

---

[24] F. W. Dyson (1868–1939) was the eleventh Astronomer Royal at Greenwich from 1910 to 1933; he is known today mostly for work in fundamental astronomical measurements and for his unusual ability for collaborative research.

[25] In 1900 Kapteyn began his famous astronomical series, the *Publications of the Astronomical Laboratory at Groningen*.

[26] See his *Plan of Selected Areas* (1906), discussed herein on pages 57–66.

relationship sometimes became strained. Consequently, their letters usually began
with a hearty greetings and excuses. Both had busy, tense lives from which their
correspondence suffered.

Simon Newcomb

In 1899 the renowned American astronomer Simon Newcomb (1835–1909) visited
Groningen. My father had already corresponded with Newcomb and had the
greatest admiration and respect for him. At the time, Newcomb occupied a unique
place in science. He was, what we here in Europe find very common among
Americans, a self-made man. His father, who was a well-traveled small-town
school master, had intended for the younger Newcomb to become a carpenter, but,
as the younger Newcomb wrote in his autobiography: "I had indeed gradually
formed from reading a vague conception of a different kind of world, a world
of light, where dwelt men who wrote books and people who knew the men who
wrote books.... I longed much to get into this world, but no possibility of doing so
presented itself."[27] By accident he came into contact with a doctor[28] who desired
to teach Newcomb the trade. The doctor, however, was a humbug behind his mask
of learnedness and affability, and the boy utterly failed to find the world of light of
which he had dreamed. Rather, he found a world of disillusionment and unfulfilled
desires, because he learned nothing and was only used as a gopher that did not
bring him closer to what he wanted. As a result, he ran away without knowing
where to go. He was even unsure if it was right or wrong what he had done.
"Am I doing right or wrong. Am I going forward to success in life or to failure
and degradation? Vainly, I tried to peer into the thick darkness of the future. No
definite idea of what success might mean could find a place in my mind. I had
sometimes indulged in daydreams, but these came not to a mind occupied as mine
on that day. And if they had, and if fancy had been allowed its wildest flights in
portraying a future, it is safe to say that the figure of an honorary Academician of
France, seated in the chair of Newton and Franklin in the Institute would not have
been in the picture."[29]

However, things went otherwise. After floundering awhile, he returned to his
father who found a place for him in Maryland as a teacher's helper, that allowed
him to develop strong mathematical skills eventually leading to his position as
arithmetician at the American Nautical Almanac.[30] He was now in a position to
pursue his mathematical studies at Harvard College in Cambridge, Massachusetts.

---

[27] Newcomb, Simon: 1903, *Reminiscences of an Astronomer*, Houghton, Mifflin & Co., Boston,
p. 21.

[28] Dr Foshay; see *ibid.*, pp. 23–61.

[29] *Ibid.*, p. 45–46.

[30] The American Nautical Almanac Office was founded and located at Harvard University in
Cambridge, Massachusetts. In 1866 it was moved to Washington, D.C. See Moyer, Albert: 1992,
A Scientist's Voice in American Culture: Simon Newcomb and the Rhetoric of Scientific Method,
University of California Press, Berkeley.

At Harvard he distinguished himself so that in 1861, his twenty-sixth year, he was named professor of mathematics in the United States Navy and astronomer at the Naval Observatory.[31] As a result, he became a 'Denizen in the world of light'.

Later he was offered the position of director of the Harvard College Observatory, which he turned down. "Perhaps unwisely for myself, though no one who knows what the Cambridge Observatory has become under Prof. [Edward] Pickering[32] can feel that Harvard has any cause to regret my decision." About the same time, he was named Superintendent of the Nautical Almanac, a post he held 20 years until he retired. In 1884, he became professor of mathematics and astronomy at The Johns Hopkins University. Numerous distinctions were bestowed upon him, including the great Huygens Medal given by the University of Leiden once every twenty years.[33]

Newcomb produced numerous astronomical works, but, in contrast to most astronomers who only concentrated on their own particular topics, he was also unusually versatile. As the English astronomer Sir Robert Ball[34] put it, "Newcomb is as versatile as he is profound." Newcomb was president of a variety of scientific organizations in which he held an interest. He possessed a great interest in psychical research, and as president of the American Society of Psychical Research he dedicated much of his attention to spiritualism. After personal experiences, however, he deemed all this as mere illusion. He kept himself extremely busy with a variety of financial projects, and even made a name for himself as a novelist.

Planning his first visit to Groningen with father in 1899, father became somewhat concerned, because he feared that he would not have much of importance to show Newcomb. Thus when there was a question whether or not Newcomb would really be coming to Groningen, father wrote him: "You will not see my instruments in this case, but this will only save you so much disappointment, as there is certainly no observatory in Europe, not to speak of America, so scantily equipped."

Still Newcomb came to Groningen and he appeared to be a remarkably stern man who seemed imposing because of his physical persona. Despite Newcomb's appreciation of my father and how much he learned and seemed to be interested

---

[31] Following his commission as professor of mathematics in 1861, Newcomb remained at the United States Naval Observatory in Washington, D.C., until 1877 when he was appointed Superintendent of the Nautical Almanac Office remaining until his retirement in 1897. For additional details, see H. Plotkin, "Astronomers versus the Navy: The Revolt of American Astronomers over the Management of the United States Naval Observatory, 1877–1902', *American Philosophical Society, Proceedings*, 122(6) (December 1978), 385–99.

[32] E. C. Pickering (1846–1919) became director of the Harvard College Observatory in 1876 where he remained until he died in 1919. He pioneered three main areas of astronomical research: visual photography, stellar spectroscopy, and stellar photography.

[33] The Huygens Medal is given in honor of the great seventeenth-century Dutch scientist Christiaan Huygens (1629–1695).

[34] Trained as an astronomer, Robert S. Ball (1840–1913) became known best for his many popular books on the history and development of astronomy.

Fig. 7.   Simon Newcomb, 1899. (Courtesy of United States Naval Observatory, Washington, D.C.)

by everything, father did not feel at ease around him but was rather prodded by Newcomb's visit. It was quite the opposite for our mother, however. She had an easy, unaffected manner in dealing with greatness that was at the same time calming and easing as well as great in and of itself. She did not waste time thinking of the impression she was making but was only interested in making her guests feel comfortable and at ease. It was not long before she was charmed by him. The fact that he was so quiet did not bother her much either, because she had enough to say and she spoke perfect English. "He is a king among men," she was fond of saying. And how Newcomb would surprise us children when he would stretch his tired, dirty feet on the beautiful golden stools that were the holy of holies in mother's salon. She baked buckwheat cakes for him for breakfast, because he loved them so. She was the friendliest hostess that a stern American could imagine.

The feelings became mutual. Newcomb later wrote Gill in 1899: "I have read a letter of yours to Mrs. Kapteyn [and I admit] it is hard to keep anything from so delightful a woman." Gill subsequently wrote my parents about Newcomb's personal observations, which thoroughly pleased them. Newcomb had a rather unusually dry sense of humor, which for such an imposing man surprised us children. He was big and robust, with thick wavy white hair that was rather nice looking. He told them that his hair was the pride of his life. Newcomb told us children that in the lexicon his name shall be known as, "The man who had the most beautiful head of hair. Seems to have been an astronomer."

Because of a malady of the nerves in his leg, Newcomb temporarily needed crutches, which only made him a more imposing figure in our eyes. He did not bother us though, which seemed quite natural to us, and so we behaved ourselves and were very quiet in his commanding presence. Still, he was very shy, and had difficulty expressing himself. His appreciation and interest expressed itself in a dry and yet touching way. Later in 1907 he wrote father: "Next year I pass through the Hague on my return from Rome. If I do this, I hope you will be able to place yourself near my line of movement." And later that year: "I will spend a week or two at the Hague. Perhaps you and Mrs. Kapteyn can also come to that region which I believe is very pleasant in the early autumn, when I shall probably be there." Once he wrote father, "I think it is nearly a year since I have heard from you personally. And I now write rather from a general desire to hear how you are doing than from having anything important to say." These few words from this man can be regarded as remarkable.

Their first visit, which was followed by others, was of great importance to father and a stimulus to his cosmological studies regarding the architecture of the stellar system. While the subject had kept him occupied for so long, still it had taken a back seat to his work with Gill. Fortunately, the cosmological problem increasingly began to involve him more and more. Many letters from this period concern these investigations and when the *Cape Photographic Durchmusterung* was finished he was able to invest his full strength on these great problems again.

The Two Perfect Gentlemen

In the year 1900 Sir David Gill visited Groningen for the third time. He was welcomed as a tried and true friend. The Boer War[35], which had caused much indignation among the Dutch, was in its last phase. As result, the English were not very popular at the time. Their greedy and unfair politics made them hated by the Hollanders, who chose enthusiastically for their own kindred in that far off land. Our family had decided not to discuss this most troublesome topic with our English guest and so we kept quiet about it. The visit by Gill was just as nice and refreshing as always. His hearty interest and spontaneous, lively spirit was of great influence on my father's work. Everything he said was stirring and to the point, and his philosophy was a help and comfort in these times of trouble. His warm friendship was benevolent and his humor irresistible. He let himself be photographed with my sister and me, whom he called his 'Lassies' and whom he teased and spoiled and was the funniest uncle that anyone could imagine. Gill and father visited the physical laboratory at Groningen where above every door a wise saying was painted. One said, "Wisdom is better than rubies." "I know what that means," Gill called out: "Whisky is better than red wine!" He had as much fun with his joke as others. Typical of Gill was his humorous rending of the Latin saying: "'Experientia docet' – 'you know: experience does it'."

"I've learned all my life's philosophy from Gill," father was fond of saying. In his letter to Gill on the latter's seventieth birthday, father wrote: "I think I picked up something of your great 'Lebensweisheit', of your capacity of making life a joy to yourself and to others." There was certainly much truth in this, for he acquired many hints of understanding from this learned and wise man that came to him just when he needed them most. Father's own wisdom, however, was not any less: His inner life was a serene surety of the soul. But there exuded from father a quiet strength that many could not understand nor explain, because, in contrast to our time, he was so simple and without pretension. Following his death, one of his clerks, who worked with him for twenty years at the Laboratory, said: "Although he had so many deep philosophical beliefs that were expressed so simply, sometimes they went over peoples' heads. But I always realized the wisdom of it. He has given my life stability, and, in difficult moments, if I thought of him, he was a help to me." This clerk told me much more of father: "Perhaps you find it irreverent that I continue to speak of 'he'. But that isn't it, it was always with a capital 'H' for me." Consider the old wise saying: Simplicity is the trademark of truth, and used everyday without thinking, we hardly feel the deepness of the truth anymore. More so than ever, father's being illustrates this in all its beauty.

After Gill returned to the Cape, their correspondence commenced. Writing letters was difficult for my father, because he was so busy and over-worked. In

---

[35] Also called the South African War, the Boer or Anglo-Boer War (October 1899–May 1902) was fought between Great Britain and two Boer republics – South African Republic (Transvaal) and the Orange Free State. The Boers are of Dutch and Huguenot descent.

answer to a very pressing letter from Gill, father answered: "The fact is that I am and have always been a bad correspondent, wanting some direct stimulus for writing a letter. Add to this the feeling that I had better avoid writing on the subject of this disastrous war, a subject which as soon as we Dutchmen write to an Englishman will come uppermost in our mind, and you will see, how it is that I don't let you hear for ever so long."

Gill was touched by this expression: "You certainly are a bad correspondent, but there is this about you, when you do write a letter, there is always something or rather there are many things well worth reading in it. Now first of all I really did not realize fully before, what a perfect gentleman you are – I did not realize till now that you feel so strongly about this miserable [Boer] war. I only wonder, how you had the power to keep so completely away from the subject during the time I was with you in Holland. As your guest of course I did not open the subject [was he also not a perfect gentleman?][36], but if I had felt as keenly as you do about it, I do not think I could have refrained so perfectly as you did. I only feel about the war that we had to fight or to make up our minds to submit to a Boer Republic throughout South Africa. But I have no feeling of animosity against the Boers. Blood is thicker than water, so if you really want to let off steam about the war, you will find me an excellent safety valve, and in the future don't let the fear of speaking your mind interfere with writing to me on that or any other subject. Indeed it would interest me greatly, if you did so." And so this friendship continually became stronger and more complete.

---

[36] Added by H. Hertzsprung-Kapteyn in the original.

CHAPTER 4

UNIVERSITY LIFE

Life in Vries

In the mean time, we children grew up. My sister and I went to the boys' Hogere Burger School and the gymnasium, which was unusual in those days. My father did not distinguish between my sister, myself, or my brother, who also went to the H.B.S. We all had the same opportunities and each of us had the same rights; so it was and so it remained for him. Father had the same interests in problems regardless of gender, and he felt deeply about the future of all three of his children. He wanted all of us to have the necessary skills and education in order to prepare ourselves for whatever we might feel called to later in life. No lessons, which he considered useful to our later life, were too costly not to learn. Father and mother denied themselves any luxury for it. Actually we three children were their luxury for which everything else was secondary. Even though father was always interested in our school work, he made sure that it did not occupy too large a role in our lives. He was opposed both to excessive ambition and to the struggle for privilege, which he considered to be fatal – morally, physically, and spiritually – to a young person's life. When my older sister came to him beaming to tell him that she was number one in her class, for which she had struggled hard, his only answer was: "Don't ever sneer that in my face again!" With this response he showed a wise and deep insight into the real worth of life, and my sister learned a lesson she never forgot. That is why he was also adverse to all types of competition, and why he never wanted his children to take any part, pointing out to them the emptiness and worthlessness of the approbation.

As a glory of fatherly love that goes far above the measured norms of child-raising, there was one moment from among all the others that will never be forgotten. There was once a large bunch of grapes on the table, which itself was very rare in such a simple family. As a small child, I eyed them gleefully and sighed: "Oh! If only I could eat that whole bunch alone, then I would be perfectly happy!" No, responded the righteous pedagogue, that would be egotistical – all must share. Isn't this true? However, he said: "Well now, then you shall enjoy perfect happiness this once child, which is so seldom." And so he gave her the whole bunch, which amazingly the others did not begrudge her. Although the feeling of complete happiness has long passed, the memory of the love of this father who wanted to bring happiness where he could has remained.

He not only had an interest in his children, which is often the case with most parents, but other children also enjoyed speaking to him and telling him of things they felt were important. For example, once there was a small girl who at school had learned that the earth circled the sun, something that she simply could not understand. When she asked her father for an explanation, he said: "Ask Professor Kapteyn, he knows better than I do." So she went to my father, and he explained the idea so clearly that she came home completely content and totally convinced. Many parents would ask my father his advise on how to raise their sons or which trade they should choose, and so on. They always left with the right advice and with a warm feeling in their hearts.

Father was interested in everything that lived and struggled in life since all were alike. He taught his children to have respect for the opinions of others, and not to respond so quickly with criticism. He was himself the best example of this. Consequently, how surprised we were when once we heard him sharply criticizing one of his colleagues. We were so used to his humane and understanding judgement that it made a deep impression on us. He understood that criticism has its value and that it does not miss in its efficacy.

At great cost he bought a summer house in the city of Vries in the north of the province of Drenthe only a few hours from Groningen. Here, with a garden, the family could spend their three summer months. While my sister and I did our housework, mother worked the whole day in the vegetable garden where her heart lay. Many guests visited, and everyone was always welcome in our home where there was always a simple and pleasant hospitality. There was a big school room where in the morning anyone who had a desire could study, and where father always spent his morning hours because his work never left him alone. In the afternoons he rode about the area on his bike in order to get to know everything. He had an open eye, an open ear, and an open heart for everything in nature. He listened to the birds and soon learned their songs. It became a dear love of his and his face would light when, unexpectedly, he heard a bird, which he would then silently approach to observe. Bird excursions lasting an entire day with binoculars and an inexhaustible enthusiasm were a joy for him. No problem or effort was too much. Through ditches and canals through thick weeds he went if he had a goal in mind. Professor Boisevain, who often accompanied him on his outings, once decided he had a bit too much: "But," he later related, "I ran after him and I thought: where he can go I can go also." Under such enthusiastic leadership the impossible became possible.

And then to see the gamecocks dance was the best of the best of a happy event in nature that father could think of. Thysse, our first fowler, described it this way:

> Everyone carries its own colors just like the shutters on a house from the Middle Ages. Here is one with a blinding white comb and pearl-gray, bordered-white feathers on its back. There is one completely in orange, and still another that looks like it got wallpaper on it with magpie feathers colored steel blue. Another one has stripes of red and blue, and across from him bows a knight with an ermine comb full of black spots. And there is never any rust in all of those feathers. It appears as if the wind repeatedly blows through the feathers,

but then comes from different directions simultaneously. And over there are the foxtails standing motionless in the morning sun. The birds themselves spread their feathers and fan them and fold them again just as a dancer does her dress. Look how they turn around their thin pole legs! Three pass in a row, bow to the ground, stand up, now with head in the neck, then beak in the air, the neck stretched out so far as possible. In tripping steps they back up against the flanks, now again the normal position and a shake of all the feathers like a poodle just coming out of the water.

Father was delighted to view this festival of color and movement. Out of the sack early, long rides on unpaved roads with endless patience, he came back home with the light of a great event in his eyes and an ever growing surprise and respect for nature.

The stones also began to speak to him. His long geological hammer became his true-blue friend on his rides over meadow and sand. Heaps of stones and grooves of earlier epochs all became the wonder of nature. Often the children were very impatient when he stopped at every pile of stones during their joint bike trips, because they did not yet realize the calling of science that gave everything such a gleam. But his drive never diminished and, after awhile, the children also searched and pounded on them with as much joy as they could for a fossil or an interesting stone that would enrich their own collection. They read and discussed geological books that gave them new ideas.

Once father discovered a few hills near Vries that, because of their frequency of appearance, made the impression on him that these had probably been man-made, ancient grave-mounds. For a few guilders, he purchased from the land owner one of hills and, with pickaxes and shovels, the whole family accompanied by friends went out with him to attack the hill and discover its secrets. Although there was much digging and working, nothing of importance was ever found. Even though the diligence began to wane and the project was eventually given up, the study of geology remained. The scientific interest was rekindled again in a new form when nature spoke of her miracles and we wanted to penetrate deeper into her mysteries. With this in mind father asked the geologist G. A. F. Molengraaff[1], who made annual excursions with his students, if he would join us. So a few times he joined as the youngest and most enthusiastic of Molengraaff's students.

Hendrik Brouwer[2], also a geologist, remembered father this way: "Because of his simple friendliness, he eventually drew all the young men to him. He never tired on long marches. Even when others were soaking wet from the rain, he was one of the few whose humor never changed." One student reminisced: "Kapteyn often asked the leader for a clearer explanation, motivated by his interest in the general results of geology. Although we usually lost track of the details ourselves, it was still interesting to listen to these discussions. Vividly I remember dinner

[1] G. A. F. Molengraaff (1860–1942) was professor of geology at the Technical University at Delft.
[2] The Dutchman H. A. Brouwer (1886–1973) was professor of historical geology and paleontology at the Technical University of Delft (1918–1928) and at the University of Amsterdam (1928–1957).

on the last day of the excursion when we usually held many speeches, and where Kapteyn spoke directing himself to the students, encouraging us to be happy of heart and not dwell upon the black, shadowy side of things. It was all very plainly said and everyone felt that Kapteyn had given something to us, all of which made a strong impression." Although it was long ago and the student has long since became a man of importance in the world of science, father's memory remained unblotted just as with everyone else who came in contact with him. A sensitive and a great man father remained in the simplicity of his words.

The Professor

It is interesting how Pieter van Rhijn described father's lectures: "If the students tried to show off their learning, Kapteyn used to say: All that is good, but I would really like that these gentlemen would see through the problem, and look at the questions more in the way of physics than of mathematics." Then followed what father believed; an explanation so transparent and understandable and so deliciously plastic. "That's it," said van Rhijn, "for which your students are especially thankful, that you have freed them from their scholarships." Continued van Rhijn:

> He did not demand much positive knowledge from his students, and in his lectures he used few facts. Rather he showed his students how things that we see as unchanging are problems in science, and how earlier generations handled and resolved them. This was the secret of his fascinating lectures. One got the impression, not that the Professor spoke and the students listened, but that the Professor and the students worked together for the resolution of a scientific problem.

His lectures were thorough, clear, and well-organized. The goals of research and theory were very clear to his students. "I have the most beautiful scientific research," wrote van Rhijn's student Jan Oort[3]. "In his company," continued Oort, "at the lectures there was something unusually arousing. Often I have come from there with more beautiful and happy thoughts than those with which I had arrived. Like no one else, Kapteyn saw the beauty of nature and science that he was able to connect so strongly. Just as the other lectures made me lose my self-confidence during my first year, the astronomy lectures helped bring it back and build it up." A better opinion could not be offered of a teacher, and it is highly characteristic of father's being.

Father did not like the teaching methods at the time. Cramming for exams and the intense erudition did not seem to be the way to create independent, thoughtful people. The most learned often stood strangely and clumsily outside reality, and often they could not solve the simplest of problems. To confirm what the teacher

---

[3] Jan H. Oort (1900–1992), who was van Rhijn's student adopting Kapteyn's and his mentor's cosmological interests, became director of the Leiden Observatory (1935), and is one of the twentieth-centuries most influential astronomers.

means one should use examples from experience. As a supervisor for many years, father took part in the gymnasium final exams. Often he had an opinion different than the other teachers concerning the mathematical ability of a candidate. For example, once there was a boy who had solved a problem involving interest on a loan. But because of a small mistake the answer became so formidable that he had to put the paper horizontally to find space to write all the numbers. The teacher found the problem correct and the small mistake of no importance. Father, however, thought that the young man should not have passed the exam, because it showed that the boy had no sense of what the problem was about; as a result, the boy completely lacked an understanding of the mathematics.

On the other hand he frequently gave students, who other teachers had judged insufficiently capable, another chance to make up for their apparent lack of understanding by demonstrating a knowledge of mathematics that father considered to be much more important. On other issues he also felt the same. For example, there was a candidate who had to render a French translation about a tired pilgrim in the desert "Comme son cœur rit, quand il s'approche d'un gîte." Obviously he didn't know the word 'gîte' and translated it as "How his heart laughs when he sees a wild animal," probably thinking of 'gibier' ('animal game'). A peculiar pilgrim indeed! Father felt that it would be better to leave it untranslated than to use such illogical absurdities, which showed a lack of common sense. "Give me the failed candidates," father would say, "and I will turn them into successful ones which you will get as a present from me."

For more than forty years he was Professor at the University of Groningen, where he remained until his retirement. He rejected nominations to the observatories at Utrecht and later at Leiden. Because he loved the observatory at Leiden so very much, the decision not to accept the Leiden position after the resignation of the astronomer van de Sande Bakhuyzen was very difficult. But he simply could not leave his astronomical laboratory, that beautiful creation of his spirit.

The Versatile Life

Every morning by 9 o'clock the residents of Groningen would see father on his way to his laboratory. Everyone knew him by his characteristically boyish way of walking with small shuffling steps, which he called the 'meadow-step' and that he attributed to his many walks as a young boy through the meadows. Before noon no one was permitted to disturb him, and in those morning hours of concentrated effort at his writing table his spirit could fly away from all other cares. He had a wonderful ability of concentration; nothing could distract him. At home he worked best in the family room where the warm atmosphere of his family stimulated him.

"I learned how to concentrate in my youth," he said laughingly. "Once when the chatter of the birds hindered me at my work, I shut the window and my mother got mad at me for my foolish sensitivity. She boxed my ears and threw open the window. That slap was more useful to me than many well-meaning words."

Also, working in the middle of a lot of boys taught him to shut out the outside world whenever it was necessary. The flip-side of his intense concentration was his tendency to absent mindedness; fortunately, he had caring people around him to help. "Professor, you haven't got your hat on yet," reminded his clerks whenever he was leaving the laboratory; or "your pant-cuff is turned up." Sometimes they would need to telephone to remind him of an appointment. His wonderful memory, however, was of immeasurable worth in memorizing exams and other important facts of the day. For example, once while enroute to America he noticed he had forgotten his wallet when he reached Rotterdam. He had just enough time to go back to Groningen and return again in time to catch the boat. Once again while on the train in America he lost his wallet; after looking high and low, the conductor was consulted. "Just look under your pillow, Sir," was his counsel. Just as he said, there lay the wallet.

The clerks at the laboratory, whose numbers continued to grow, were very attached to him. A respectful love and an unshakable faith filled them. No wonder, for father shared in their joys and personal sorrows, supported and counselled them when they needed it most, had an unending trust in them, and, as they all felt, was altogether like a father to them. In his laboratory there prevailed an orderly atmosphere that father brought with him whenever he was present and that lightened the work-load that often was endless, routine, and very boring. His beaming optimism helped them through many difficult times. "Come, let's first begin with a manageable size." Usually when the problem was tackled with such courage, it solved itself.

Afternoons were set aside for contact with the world outside his laboratory: colleagues, lectures, exams, and his walks. Especially his regular Monday afternoon walks with his two trusted friends, G. Heymans[4] and Boissevain[5], were important. The three were a well-known sight on the path to Haren path every Monday.

Usually in the middle of the three was Heymans, whose imposing figure in the palerine jacket with the Paraway and whose lofty spirit rose above the rest of humanity. In contrast, Boissevain was very small, always moving, and full of interest in life. On the other side was father, slender and animated, enjoying being outside with his interest in everything, and who was always identifying every bird call. As with a ritual, everyday for twenty years they walked to Haren, a village about an hour from Groningen. It was remarkable, they concluded, how seldom people were forced to give up their walks due to the notorious Dutch weather. In later years, they would end their walks at the Cafe de Passage, where they would sit and talk awhile before returning home pleased with themselves. It was a beautiful friendship full of mutual respect and appreciation.

Father suffered a great deal from a nervous pain in his eye for which he spent

---

[4] G. Heymans (1857–1930) was a philosophically inclined psychologist at the University of Groningen (1890–1926) and, while at the University, he established the first psychological laboratory in Holland in 1893.

[5] Boissevain was a professor of history at the University of Groningen.

years looking for a cure until it eventually healed itself. Because of this malady, for a long time he felt compelled to give up his evening readings. This was a great sacrifice but good friends helped him and read to him aloud on a regular basis. For many years Heymans did this and both enjoyed it. At first they read articles in father's field of study; later other things that interested both of them, primarily articles or advertisements of contemporary scientific problems. They discussed the problems, but because they were so different in their backgrounds and temperaments, there was much discussion that always raised new questions and put things into a new perspective. Both had great respect for science and a solid belief in its sovereignty.

Typical of my father was his strong feeling of scientific honesty. For example, in 1908 Frederick Cook[6] maintained that he had reached the North Pole. Later, however, it appeared that he had actually lied about his earlier alleged assent (1906) of Mt. McKinley. Many had been quite content with the light-handed way the scientific world had accepted Cook's deception. My father, however, came down hard. Although he accepted it as natural that people would believe a scientist's word and would not think at all of the possibility of having been deceived, the hypocrisy inflamed father's indignation. Even after many years his anger became aroused when a naturalist, who had developed a particular theory, asked father to read certain books to substantiate his theory. The naturalist urged father: "You don't expect me to gather facts against my own theory!" Regarding another well-known biologist in whose writings he had noticed evidence of deception, father said: "That is the only man who I hate."

Twice in his life he had come in contact with men who were not honest in their science. In this he was immovable, and it was as impossible to change his attitude as his own spirit. In his view these scientists were forever lost having fallen in the most basic demand of scientific honor; they were no longer worthy to carry the scientific torch.

Heymans and father studied other works as well. The famous *Novum Organum* (1620) by Francis Bacon (1561–1626) and *The Life of Samuel Johnson* (1791) by James Boswell (1740–1795), that bulky volume that everyone knows by name but that almost no one reads. Together they studied Einstein's 'theory of relativity', which interested my father but that he "couldn't see through". The physicist Paul Ehrenfest[7] came to Groningen to lecture for a few days to discuss it. They also discussed philosophical articles; but father's strong sense of scientific realism prevented him from enjoying them, because he had difficulty "getting into them". One exception was an article that had formed an exact psychological investigation

---

[6] The American physician and explorer F. A. Cook's (1865–1940) claims were successfully challenged by Robert E. Peary, who is credited with being the first person to reach the North Pole in 1909.

[7] The Austrian P. Ehrenfest (1880–1933) became professor of theoretical physics at the University of Leiden in 1912 where he remained until his death in 1933. Ehrenfest was known as a brilliant teacher, who made novel important contributions in statistical mechanics and in quantum mechanics.

for which he was always interested in. With great care he followed Heyman's own personal research in that area and was often able to help with the mathematical formulations.

His great desire to help was one of his most identifying qualities. For example, some of his writings on statistics were among those written with the help of M. J. van Uven, who later became a biologist at Groningen.[8] Gill believed, however, that all this occupied far too much of my father's time, because father was an astronomer at heart and soul. Therefore, in 1907 he wrote to my father:

> Why a man of special astronomical gifts like yourself should waste his days in abstract mathematical work which so many men are capable of working at – whilst there are so few to do what you can so well do – I don't know. After all what is the value or interest of a frequency curve compared with the structure of the universe? I am glad to hear that you confess to a temporary possession by an evil spirit – some form of exorcism is necessary – and I wish to administer it, if I can. I do think that in astronomy at the present time there is nothing comparable in interest with your work and the plans you propose for its accomplishment.

Still, he was willing to take up correspondence with teachers and others who imagined to have discovered errors in Newton or other meaningful discoveries that always led to nothing, but still never made him impatient.

Despite the fact that father did not feel much for pure philosophy, because his nature did not reflect it, still he went to great lengths to stay informed. For a year he followed the philosopher Bolland[9] not because the personality and teachings of the man attracted my father – in fact they repelled him actually, appearing to him unbalanced and undisciplined – but because he wanted to understand what so many others could not. Sometimes Bolland's fiery way with words would lift him. But all too soon his remarks would become less motivating and illogical, and his reckless disparaging remarks of others who thought differently awakened father's indignation. Although he was level-headed, his hot-headed and undisciplined temperament made my father feel that his absolute subjectivity was just as foreign as an unsympathetic one.

According to my father and as he had become used to thinking since a youngster, science must be thought of in terms of natural laws; above all one must be objective. Without objectivity father maintained that one would become as helpless as a ship without ores, unable to expect anything good to pass or like becoming lost as in a fog. Father found Lessing's philosophy and classical tranquility appealing; in Lessing wisdom was kindred. My father contributed to his philosophical mentor (Lessing) whenever he hoped to add more to literature by further studying this philosopher. A few days before his death, my father asked for Lessing's complete

---

[8] Kapteyn, J. C.: 1903, 'Skew frequency curves in biology and statistics', Noordhoff, Groningen, 45 pages, and Kapteyn, J. C. and van Uven, M. J.: 1916, 'Skew Frequency-curves in biology and statistics. Second paper', Astronomical Laboratory, Groningen, 69 pages.

[9] Bolland was a professor of philosophy at the University of Leiden.

works. He could only see the bindings and rejoiced at the material; alas, they remained unread. Many things remained undone that my father had promised himself for his time of rest.

Hudig, a young neighbor and also a good friend, read to him regularly from geological and historical works; those evenings were a joy for both of them. My mother also read daily to her husband from novels they both chose, as well as from flowery and colorful magazine stories that amused father. They worked a lot together, because more and more it was mother who wrote father's letters as he dictated them. Because his correspondence was extensive, her help was of immeasurable worth. And so went the years, until his eye eventually healed itself and once again my father became master of his own lot.

For many years father was president of the Scientific Chapter of the Physics Society of Groningen whose primary aim was for its members to stay abreast of science through lectures. Speakers who were foremost in their fields were invited to come and share the results of their investigations. Along with Hudig his energetic secretary, my father knew how to attract many scientists to come and speak. Among those invited were the anthropologist Eugene Dubois[10] and the ornithologist Dr Tienemann[11], whose interesting work appealed greatly to my father the bird watcher.

During father's presidency the Society reached its zenith. My father held many of the readings himself and was tireless in helping to keep the scientific community in Groningen at a high level. After the readings came the most important part: the gathering at the Restaurant Willems, where everyone could gather and talk further about the interesting things they had heard earlier in order to share their insights.

Even though father wasn't musical, he didn't play an instrument and he didn't have a nice voice, around his sixtieth year he began music lessons so as better to understand this art. At home he had his own way of musical expression, which was called 'trompet'. Whenever it manifested itself, mother was irresistibly driven to the piano to lead them, which made a nice original effect. People always heard him singing on his way home as well as at work. My father had been known to break out into a song, usually pieces of a known sonata or symphony, that were tried and true. He preferred to hear the old well-known parts, while the new unknown ones were not of much interest to him. Still, the art of music was a mystery to him. It filled him with a begrudging respect, and he was filled with the deepest awe for the artist. He followed a musical course given by Peter van Anrooy[12], which enormously interested him. Every Wednesday he delved into the purity of music and felt himself the richer for the experience. His meeting with van Anrooy, which quickly developed into a warm friendship, brought him closer to the art. He found

---

[10] French anthropologist E. Dubois (1858–1941) was professor at the University of Amsterdam and the discoverer of *Pithecanthropus erectus*.

[11] Tienemann was director of the Vogelwarte at Rossiten in the Kurische Nehrung.

[12] P. van Anrooy (1879–1954) was appointed director of the orchestra at Groningen (1905), Arnhem (1910), and Residentie-Orhest at The Hague (1917).

numerous parallels between science and art. In their highest form were not both unselfish, reaching for an ideal that separately they had established while seeking for truth and the purest expression? Oblivious to the fame and prosperity of the world, both achieved the highest that dwelt in them. In this way father viewed art as the sister of science.

Poetry, however, remained a mystery for my father. Not that he looked down on poetry with compassion, as so many intellectuals do. It just didn't make sense to him. Judge for oneself the only poem by his hand. It was his 'Green' poem. One must admit that the youth of a young person does not lend itself well to poetic flight, but his verse was undoubtedly the most unpoetic rhyme of his contemporaries:

Het groenenvers is een eerstvereischte
Voor groenen, zoo merk ik.
Het is dus zeker wel het wijste
Dat ik mij naar die gewoonte schik.[13]

Can it be more unpoetic? Still it is interesting, for his spirit speaks with the character of his logic and philosophy!

Only epic poems could move him. His favorite was the poem of Waltharius[14], that Germanic poem of delightful primitiveness of fighting and blood, of primitive people with primitive instincts. It grabbed him and awakened his enthusiasm. Inconceivably many saw this in him, but it was precisely that bit of simplicity, that bit of primitive nature, which was unquestionably his own, that found freedom and expression in this thundering song of supermen it its unbridled passion.

In many ways the purity of nature revealed herself to my father; he simply had an unending thirst to learn and enjoy from it. Each spring the first flowers and birds became a great joy for him. From his walks he would bring the first daisies, which he would lay on the table half shyly and half joyfully. Every new bird call would fill him with rapture, and the appearance of the first swallows were noted on the calendar. "In my next incarnation I would like to be a swallow," he often remarked. To him they seemed the personification of an unworried and joyful life. Another time, however, he wished to become a dandy. The beauty that derives from a neat appearance always impressed him. "There is much self-confidence in a nice appearance," he often said. Although he never reached this ideal, the highest that he attained was an ordinary clean appearance.

He could laugh, full and heartily, especially uncontrollably if he told an anecdote – whether it was the first time or the twentieth it did not matter. He laughed until the tears streamed from his eyes. The most sober among his listeners was always won over by this gift of true joyfulness of heart, and even forgot his darkest of thoughts and laughed as well. A wise Frenchman once said these meaningful words: "The sage does not laugh, he smiles." My father knew better. In his

---

[13] "The green poem is highly demanded / For green, so live noticed. / It is thus the most wise / That I accept this custom.

[14] A Latin heroic poem of the ninth or tenth century dealing with a Germanic hero legend written by a Bavarian named Geraldus.

'Personal Remembrances of J. C. Kapteyn', the journalist-astronomer Cornelis Easton[15] remembered my father this way: "This unusual man was even able to neutralize the oppressive influence of a new street in Rotterdam under a led-heavy, wet summer sky by way of his easy going, completely natural cheerfulness."

Neither did my father pass judgement on the cinema as those in intellectual circles liked to do. He went gladly with an open-mind, seeking relaxation without asking deeper questions. Why all the destructive critique of a hopeless triviality of life and inferiority? Why must people cling desperately to a life above a certain critique that destroys everything it touches? Anyway, just as they do in theater and literature, people can ignore the melodramatic and sensational. Instead, they should enjoy the unlimited technical and artistic possibilities that modern art offers.

For example, once there was a carnival that had come to Groningen that pitched a tent on the Ossemarket near our house. Every evening we heard interesting sounds: pistol shots, yelling, deranged clapping. When we could no longer stand it, we had to see and share in the excitement. On the first row my father and I sat watching a foolish play of some stupid detectives who were always falling and jumping. These slimy bandits who at the most critical moments sang a long song called 'The birds of the night', continued with the craziest and most unlikely situations. We laughed and clapped at such foolishness and also sang afterwards the song of the birds of the night. Here also can be seen the godly rhythm of a wonderful, multifaceted life that can be seen for those who have an eye to see it.

Life in Groningen, however, was normally calm. With the slow passage of time, those unfamiliar with life at the Astronomical Laboratory would certainly not suspect the great excitement of the slowly emerging discoveries, some research of which had not been undertaken in this area since William Herschel a century earlier.[16] Often young people who are only aware of their own growth as the most important event fail utterly to realize the great things developing right around them. Often citing Keats, they think: "Oh for a life of emotions rather than of thought!" To which my father answered: "Child, do you think a life of thought doesn't know emotions?" This was my father's great strength! For the first time I began to realize that behind the world of numbers and instruments, of dull laboratory rooms and systematical work, something of the greatness and deep emotion of the creator of imperishable values lay hidden.

---

[15] A journalist, C. Easton (1864–1929) contributed to astronomy primarily as a public interpreter of astronomical research, mostly on the nature of the Milky Way Galaxy. On the recommendation of Kapteyn, the University of Groningen in 1903 awarded Easton an honorary doctorate in physical sciences.

[16] For nearly forty years, Kapteyn's primary research was to solve the problem – 'Construction of the Heavens' – initially outlined by William Herschel a century earlier.

CHAPTER 5

INTERNATIONAL ASTRONOMY

The 'Two Star-Streams' and the Selected Areas

My father's research was focused mostly on the structure of the stellar system. For that purpose, he had to know the density of the stars[1] in different places of the Galaxy.[2] That could only be determined as the distance of all star groups moved about the Sun. Father thought he could determine this by analyzing the movements of the stars, whose apparent brightness decreased with distance from the Earth. But this proves valid only when the movements are divided up completely randomly, without preference to any direction. Initially, his method for determining star-density was based on the premise of *random* motions. During the course of his research, however, it became evident that this assumption was not valid.[3] Its invalidity eventually caused his method to fail and his research based on stellar motions to become useless. Ironically, however, this failure turned into victory. As he continued his work, which was based on the invalid hypothesis of random movements, he made the most important scientific discovery of his life: the 'two star-streams'.[4]

The division of the movement of the stars reveals itself best when one thinks of all stars grouped together at one point in the sky from which point one needs only to imagine the motion of the stars toward the sides in the form of arrows. If these particular movements were divided randomly and only the influence of the movement of the Solar System were to be added, then the arrows moving away from the apex would be the shortest and the most sparse, being divided symmetrically right and left from this direction (see Figure 8).

In actuality, however, the movements in most parts of the sky showed a much more inconsistent division: there was no single direction of preference (directed away from the apex); rather the arrows piled-up in two different, diagonally cross-lying directions. If for every point on the sky one now drew those directions on a celestial globe in which movements were most scattered, then it would appear

---

[1] Kapteyn defined star-density as the absolute number of stars per unit volume of space.

[2] Although Kapteyn and others used the word 'universe', he was actually only dealing with the Milky Way Galaxy.

[3] See Kapteyn, J. C. and Kapteyn, W. : 1900, 'On the Distribution of Cosmic Velocities – Part I, Theory', *Groningen Publications*, No. 5, 87 pp.

[4] For a discussion of this discovery in its historical and scientific context, see Paul, E. Robert: 1993, *The Milky Way Galaxy and Statistical Cosmology, 1890–1924*, Cambridge University Press, New York, chapters 4 and 5.

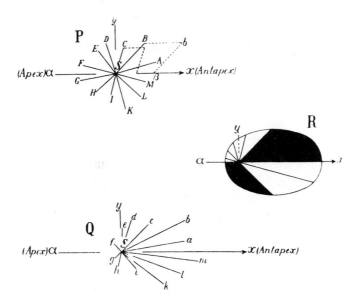

Fig. 8. 'Star-streaming'. *P* represents the distribution of the peculiar proper motions for a group of stars located at *S*. The directions of motion are randomly distributed, whereas the length of each radial vector represents the magnitude of motion. When the observer is in motion toward the apex, the observed motions of stars as in *P* becomes *Q*. In *R* the radial vectors making angles between 0° and −60° and between 60° and 180° have been blackened. (J. C. Kapteyn, 'Star-streaming', *Report of the British Association for the Advancement of Science*, section A, 1905: 257–65, p. 258.).

that they were directed towards two different converging points: One in the Orion constellation and the other in the constellation formed like a Marksman, the two about 140° separated from one another. While previously it had been thought that a single visible stream of stars moved away from the apex, it now appeared there were two streams. If the stars in the Galaxy are divided into two streams, then each stream must move in an opposite direction because the only stable point in comparison to where the movements in the Galaxy can be observed is the common center of gravity of both streams that encompasses all the stars in the Galaxy. The fact that both stream directions are not exactly contrary, but form an angle of 140°, indicates that the Solar System also has a motion with respect to the same center of gravity. Because of this, one can form convergence points. From the Sun's movement one can also find the real (and opposite) stream directions and the places in the sky to which they are directed.

For the last one, father found the following directions: right ascension 91° declination 13° north, and right ascension 271° declination 13° south. Due to this result, which Arthur S. Eddington considered one of the five most important astronomical discoveries of the century, many difficulties were removed that previously had manifested themselves in research dealing with the Sun's motion. As a result, father concluded that our Solar System could have been created by the convergence of two originally distinct groups of stars. Because not only the direction of these streams but also the movement of smaller objects (such as the constellation of the Big Bear, the Hyades, and the Perseus group) lie in the Milky Way, father expressed the view that the expanse of our Solar System in the Milky Way had been created through these stream-directions.

This discovery eventually brought about a revolution concerning astronomy's understanding of the architecture of the stars and the Solar System. But more than ever, father felt the necessity to collect data that would form the basis for further research. In the late eighteenth and early nineteenth centuries, William Herschel had built a great telescope and with it had collected more data than hitherto available.[5] Father commanded many more resources, for his plan was to get the help and cooperation of the greatest astronomical observatories in the world with their enormous instruments and means. To reach this goal, he set up a program that he wanted to submit to the judgement of some leading astronomers before presenting it to the whole astronomical community.

This was my father's famous 'Plan of Selected Area', the purpose of which was to determine for all stars within the selected stellar regions: (1) accurate photometric determinations of the sizes of the stars, (2) accurate counting of the number of stars per square degree for stars down to the fourteenth magnitude, (3) their approximate positions, (4) their stellar motions, (5) their stellar parallaxes, (6) their stellar spectra, (7) their radial velocities, and (8) the precision of the sky's base for the different parts of the viewing areas.

As the survey for the complete sky would be an enormous undertaking, father decided on a few selected areas that were equally distributed in the sky. Thus the name of the plan: 'Selected Areas'. Altogether there were 206 areas each the size of a photo-graphic plate from which all the data from the brightest to the faintest stars could be obtained.

My father first announced his discovery of the two star-streams at the St. Louis World Exhibition in September of 1904.[6] Father had been invited to give a presentation at the Exhibition, which he accepted with joy. Acquaintance with his colleagues on the American-side of the Atlantic and with their observations where beautiful work was being undertaken was very tempting to him. Consequently, he hoped to obtain cooperation for his 'Plan' with some of the American astronomers.

With mother, who accompanied him, their departure was characteristically simple. Together they left their home in Vries by bike in August and rode to the

[5] Although Herschel built a 40-foot telescope with a mirror 58 inches in diameter, this instrument was too unwieldy. For all his important work on planetary and stellar studies, Herschel used a 20-foot telescope with a 19-inch aperture.

[6] Kapteyn, J. C.: 1904, 'Statistical Methods in Stellar Astronomy', *International Congress of Arts and Sciences, St. Louis* 4, 396–425. For the broader context, see Sopka, Katherine R. (ed.) and comp.: 1986, *Physics for a New Century: Papers Presented at the 1904 St. Louis Congress*, American Institute of Physics, Los Angeles.

Fig. 9.   George Ellery Hale, ca. 1910.

train-station in Groningen from where they travelled to Rotterdam. The children
lined-up along the gate and waved good-bye on their way. Thus began their first big
journey to America. My father would go to conquer the astronomical world, and
my mother to make numerous friends from whom she would find kindred spirits.

Arriving in St. Louis they were welcomed by the Dutch Consul, who, although
the rest was fully acceptable, did not provide too dignified facilities but rather
primitive, wooden buildings. They arrived just in time to attend the opening of
the astronomical congress, where they sat themselves very quietly in the back of
the room. Simon Newcomb, the president of the astronomical section, noticed

them and escorted Kapteyn immediately to the committee table where all the esteemed gentlemen astronomers were assembled. Father was greeted with much hospitality and respect, and immediately felt admitted as a welcomed and honored guest. Newcomb, who knew nothing of his new discovery, introduced my father: "Professor Kapteyn will tell us something about his interesting Durchmusterung work." Although the audience was indeed surprised when the lecture concerned a completely different topic, still it was well received and father was satisfied with the interest shown in his discovery of the two star-streams.

The most important fact of his American trip, however, was his acquaintanceship with George Ellery Hale, then director of the Yerkes Observatory. Simultaneously in St. Louis, an international conference on solar research with Hale as president was taking place for which father and Willem Julius[7] were the Dutch delegates. Here my father explained his 'Plan of Selected Areas' to Hale in greater detail. The meeting took place on the exhibition grounds in the Tyrolean 'alps', one of America's most grandiose exhibits of artificial scenery. They sat there the entire afternoon absorbed in their discussion, and had no concern for the natural beauty around them. Hale was enthusiastic, which would later show in his enormous support of father's 'Plan'. Father was deeply impressed with the broad-mindedness and fine spirit of this American, who would become leader of the Mount Wilson Solar Observatory in Pasadena, California, which was soon to reach completion and become the biggest and best equipped astronomical observatory in the world.[8]

Hale later wrote about this first meeting with my father in his 'Recollections of Kapteyn':

As chairman of the Academy's Committee, I was deeply interested in the possibilities of international cooperation, and convinced that much might be accomplished through joint effort. I was therefore greatly impressed by Kapteyn's scheme of Selected Areas, which he presented at the International Scientific Congress, held in Conjunction with the St. Louis Exposition. My own experience had been in the field of astrophysical research, and my plans for the then nascent Mount Wilson Observatory were chiefly confined to an attack upon the physical problems involved in the study of stellar evolution, based upon a thorough investigation of the sun as a typical star. Researches on the distribution of stars in space did not then enter into the scheme. However, as I listened to Kapteyn's masterly paper and realized the wide scope of his plans and the skill with which he availed himself of international cooperation in assuring their execution, I was deeply impressed by his appeal. Could we not help him to secure the data needed for the fainter stars and at the same time broaden and strengthen the attack in our own problem of stellar evolution? The answer is obvious today, but at that time, approaching the subject along the

[7] The Dutch solar physicist W. H. Julius (1860–1925) was professor of physics at the University of Utrecht from 1896 until his retirement.

[8] The Mount Wilson facility became operational with completion of the then world's largest (60-inch) reflecting telescope in December 1907.

path marked out by Huggins[9], Lockyer[10] and other pioneers of solar and stellar physics, and seriously hampered by lack of funds, the case was not so clear. Nevertheless the genius of Kapteyn and the personal charm which brought him the unqualified support of astronomers the world over, convinced me at once that the Mt. Wilson Observatory ought to profit by his cooperation as soon as circumstances might permit.[11]

He goes on to write:

A powerfully creative imagination, glowing with optimism and enthusiasm is prone to set itself too vast a task. But Kapteyn, though he would gladly have measured all the stars of heaven, recognized the necessity of limiting his endeavor.

Hence the 'Plan of Selected Areas' was proposed and the successful appeal for international cooperation was subsequently obtained.

On their way back, my parents visited Simon Newcomb in Washington, where, at the same time, they attended a big reception at the White House. They arrived in Newcomb's carriage pulled by two horses and were introduced to President Theodore Roosevelt together with Hugo de Vries[12], who had also spent time in Washington during the summer of 1904 and 1906. The President had a friendly word and a handshake for them all. "Ah, Mr. de Vries, I suppose we are cousins, because the name de Vries is in my family." The latter, of course, was said with pride as Dutch names had distinction in America. Father listened to a friendly remark about astronomy, and then that special moment too belonged to the past.

In February of 1905, the 'Plan' was sent to 25 astronomers who were "best in a position to cooperate in its execution." It was well received and, with little exception, all extended their cooperation to my father.

In the summer of that year, father traveled to South Africa by invitation of the British Association for the Advancement of Science, which held its large international assembly there.[13] Willem de Sitter[14], who was then father's assistant,

---

[9] W. Huggins (1824–1910) was a pioneer astrophysicist who made spectroscopic observations of stars, comets, and novae.

[10] J. N. Lockyer (1836–1920) was a pioneer spectroscopist who discovered how to make spectroscopic observations of solar prominences in full sunlight.

[11] Hale's 'Recollections of Kapteyn' were enclosed in a letter from Hale to H. Hertzsprung-Kapteyn, 13 August 1926 (Hale).

[12] H. de Vries (1848–1935) was a plant physiologist and geneticist at the University of Amsterdam from 1878 to 1918. During the 1890s he rediscovered Mendel's laws of inheritance, which he, along with C. F. J. E. Correns (1864–1933) and E. Tschermak (1871–1962), announced independently in 1900.

[13] During the 1905 meetings of the BAAS in South Africa, Kapteyn again presented his discovery on star-streaming; see Kapteyn, J. C.: 1905, 'Star-streaming', *Report of the British Association for the Advancement of Science*, section A, 257–65.

[14] W. de Sitter (1872–1934), as one of Kapteyn's eight doctoral students, remained strongly influenced by his mentor throughout his career. In 1919 he was appointed director of the Leiden Observatory where he remained until his death. His main contributions were in celestial mechanics and in the theory of relativity applied to cosmology.

accompanied him. Other astronomers also traveled in his company, including Jöns Backlund[15], Anders Donner, Arthur Hinks[16], and Bryan Cookson[17]. They assembled frequently during their journey and for their transit they established 'The Astronomical Society of the Atlantic."[18] The main topic of all their meetings was the 'Plan of Selected Areas', of which each had brought a copy. Hinks made a record of what was suggested, and father was satisfied with the result of their discussions. Each one wanted to take part in the work. In South Africa he would win more participation for his project.

For a long time it had been his desire to see Sir David Gill, the British Astronomer Royal, in his own observatory. One learns to know a man only completely when one sees him in his workshop. With admiration, he observed the facilities that Gill had brought to fruition, and he saw his friend as the big boss, as the president of the British Association and its host. Gill, who was the radiating point of attention was both flexible and sociable, and was full of interest and untiring. However, he had no understanding of what it meant to fulfill his singular position with dignity. One of his friends told him not to forget that he was the president and therefore that he should "be on his dignity." "That is just what my brother said to me," was his reply. "Davie," said he, "you've no more dignity than a duck." And he stayed his amiable, jovial self to the joy of everyone around. Happy and content, father returned home. He had seen and admired Gill's observatory, had seen his friend in full glory, and had made great progress with his own plans.

He had become acquainted with the country, to which each Dutchman feels tied, and had seen it in all its beauty as the Association had organized some excursions that had interested him greatly. He was delighted with the luxurious plant growth; whole fields full of arums impressed him greatly. He found the Zambezi water falls (Republic of Zimbabwe) smaller than the Niagara falls, but a lot stranger. The Southern Cross, that much praised constellation, disappointed his expectations. As a whole, he thought the northern star sky more beautiful, but Scorpio in the Zenith was magnificent. He was most impressed by the tomb of Cecil Rhodes, which through its shear size made a deep impression.[19] There was only a huge, flat stone with the name Rhodes engraved on it, in the middle of the majestic desolation of

---

[15] The Swedish astronomer J.O. Backlund (1846–1916) was director of the observatory at Pulkova (Russia) from 1895 to 1916, and is known mostly for his life-long work explaining the irregular changes in the motion of Encke's comet.

[16] A. R. Hinks (1873–1945) was a lecturer in astronomy at the University of London (1913–1941).

[17] Before his untimely death, B. Cookson (1874–1909) had been trained as an astronomer.

[18] The name for this informal group was derived from one of the world's most important astronomical associations, 'The Astronomical Society of the Pacific', headquartered in California. See Bracher, Katherine: 1989, 'A Centennial History of the Astronomical Society of the Pacific', Mercury 18 (5), 3–43.

[19] Sir Cecil Rhodes (1853–1902) was an English gentleman who spent his life almost entirely in southern Africa and made a fortune in his adopted country. Today he is best known for having originally established 160 scholarships at Oxford University to be held simultaneously by two students from each state and territory of the United States of America, three from each of eighteen British colonies, and fifteen reserved for German students to be chosen by the Kaiser.

the Matoppo hills in Matabele country, an impressive memento of the submission
of the Matabele King, Lobengula. He became acquainted with some well-known
South Africaners and learned a great deal about the situation of the country after
the Boer war. In all respects, his trip was interesting and successful.

After returning to Holland, he immediately started working on the elaboration
of his 'Plan' in order to expose it to its widening group of supporters. In February
of 1906, the 'Plan of Selected Areas' was sent to the whole astronomical world
with an appendix for suggestions of those astronomers who had already offered
their cooperation.[20] As expected, cooperation was universal. In February of 1907,
when Hale too promised his cooperation, the success of the work was assured.
Father wrote Arthur Rambaut[21] at Oxford soon thereafter:

> The 'Plan' is making its way splendidly. Quite recently I got proposals from
> Hale at Mt. Wilson, which, if we get to some agreement, will nearly bring
> us to the point that everything that can be done with the present resources of
> astronomy, will be done.

Many commented on the best possible ways to execute father's 'Plan'. For example, Edward Pickering, Director of the Harvard Observatory, suggested forty-six
additional regions to research that showed peculiarities in the piling of empty space
mostly in the plane of the Milky Way. These were included in the 'Plan'. Others,
of course, had different proposals altogether.

With diplomacy and modesty father also accepted criticism to give the participants their justice and meaning, and yet quietly urged his own ideas. When a human
being has genius, character, and charm, he can move mountains. As a result, he got
everything done the way he wanted, and soon the work was in full progress. The
undertaking was so gigantic, however, that it soon became impossible for Kapteyn
to lead it alone. He suggested the organization of a committee and wrote, amongst
other things, to the astronomer Karl Friedrich Küstner[22] at Bonn:

> The cooperation for my 'Plan of Selected Areas' is now so large that respon-
> sibility is threatening to become a monstrous burden for any single person to
> assume. Gill has long since insisted that a conference be called together to
> resolve these issues. With all my power, I have declared against it. Almost all
> affairs are now in the best of hands. A conference would be too unwieldy, and
> would never provide the guidance for the 'Plan'. We have manifold experiences
> in this area.[23]

---

[20] Kapteyn, J. C.: 1906, *Plan of Selected Areas*, Groningen.
[21] A. A. Rambaut (1859–1923) was Irish Royal Astronomer (1892–1897) and later director of the
observatory at Oxford University (1897–1923).
[22] K. F. Küstner (1856–1936) was professor of astronomy (1891) at the University of Bonn.
[23] Kapteyn to K. F. Küstner, 22 July 1907 (Hale). Also see Kapteyn to G. E. Hale, 22 July 1907
(Hale), in which Kapteyn writes that both Hale and Gill approve the idea of a committee containing
many of the world's most distinguished astronomers.

Fig. 10.   Mount Wilson Solar Observatory, ca. 1910.

He thus asked David Gill, Edward Pickering, George Ellery Hale, K. F. Küstner, Karl Schwarzschild[24], Frank Dyson, and Walter S. Adams[25], all of whom represented the astronomical world at the time, to serve on a committee; something they all agreed to and accepted gladly.

For the last twenty years in all countries of the world there has been zealous activity and work on this great 'Plan' from which there has already received a significant harvest. As a result, father became well-known in America. Homage after homage came his way. But the event that tópped everything was Hale's proposal, in June of 1907, in which he asked father to visit Mount Wilson several months every year to lend his cooperation and share his leadership.

In 1905 on Mt. Wilson, the 6000-foot high mountain above Pasadena, California, the Solar Observatory, the most grandiose and best equipped observatory in the

---

[24] The German astronomer K. Schwarzschild (1873–1916) was professor at the University of Göttingen (1901–1909) and director of the Astrophysical Observatory at Potsdam from 1909 until his untimely death. He was a major contributor to stellar statistics, star-streaming, and cosmology and general relativity.

[25] The American astronomer W. S. Adams (1876–1956) was an assistant to George Ellery Hale at the Yerkes Observatory (1901–1905) and at the Mount Wilson Observatory (1905–1946) where he became director from 1923 to 1946 following Hale's retirement.

world, was opened. Andrew Carnegie[26], the well-known American philanthropist, had given money for its construction under the condition that everything would be first class. It was a miracle of perfection. For two years, Hale had taken tests to find the best climate. Money and trouble was not spared in its construction, and a 100-inch telescope – that is, a telescope with a mirror 100 inches (254 cm) in diameter – was to crown the work as a whole.[27] The smartest of astronomers became its director. It was no wonder then that my father accepted Hale's offer joyfully. He responded to Hale, who actually had some second thoughts about the cooperation, writing:

> As to such questions as the credit which each of the cooperators will get for work done, I have not the slightest fear of any disagreement and think we need lose no words about it. I never had the slightest disagreement with Gill, with whom I have worked for so many years. I felt proud to work with him for a great end. I feel proud that now again I may work with a man such as you for still greater ends.

His cooperation was now assured by Hale, and after the Carnegie Institution nominated him as Research Associate, father was able annually to spend three months at Mt. Wilson.[28]

It was peculiar that people in Holland were so little interested and knew so little of my father, who was now already known internationally. When, for instance, about this time he requested a stereo comparator from the government, the First Cabinet had reason to ask "whether the Laboratory in Groningen is not too much besides the two observatories in our country."[29] He was invited to elaborate on the reasons for the existence of his laboratory. It seems ridiculous when one understands the leading roll that father's laboratory then already played in the astronomical world. Truly great and unselfish as father was, this indifference and lack of interest did not bother him. He sought neither fame nor fortune, only the financial possibilities to continue his work. That is why he was provided with the funds and was given the stereo comparator the following year. They were most likely informed about his worldwide fame.

Home

The children were now all grown and all three were studying at the university. My sister and I were among the first female students; my sister had chosen medicine

---

[26] A Scotsman by birth and an American by choice, Andrew Carnegie (1835–1919) made his fortune in the American iron and steel industry. As a philanthropist he used his fortune to establish educational and research facilities, including the Carnegie Institution of Washington (1902).

[27] As mentioned earlier, the 60-inch was first constructed becoming operational in 1907; not until November 1917 was the 100-inch Hooker telescope finally tested becoming operational soon thereafter.

[28] Kapteyn spent each summer term between 1908 and 1914 at Mount Wilson, after which the war years permanently interrupted these annual sojourns to Pasadena.

[29] The other two operational observatories in Holland at the time were at Leiden and Utrecht.

and I had chosen law, which caused a lot of criticism in those stormy days of women's rights. But my father found it completely natural for women to study, so no one argued the point with him. My brother was studying mining engineering at the University of Freiberg in Germany (Saxony).

It was very comfortable and colorful in the great upper house on the Heerestraat where my parents displayed much hospitality. Things were simple and good there; youth, music, and song surrounded them. Plans were made for the future; young people's problems were discussed, counsel and comfort was sought and given. Father was always ready whenever required to solve a problem or when a judgement was needed. He was both the counselor and friend, who could look at things from many points-of-view and who was in all cases at home. He was different than most eminent and learned scholars, whose studies had come at the cost of the human side of life. "What a remarkable person," a simple woman once said with a certain shyness when she had visited our family. "He was so ordinary and truly had an interest in me, who means so little really." Every person meant something to him. No one ever got the feeling of pressure or of little worth, which great people can often make one feel.

In 1905 there was a revolution in our house. When my older sister got married and, while I was still studying in Amsterdam and my brother engineering in Freiberg, the house became too big for them so they decided to find a smaller one. Mother, who was on her way to America feeling great sympathy with her American sisters and their regiment of self-help, prevailed upon my father to break with the Dutch tradition of having a maid and a working woman for the morning hours. She was the first in her circle to make this bold move and her husband found it excellent, because it appeared that her work didn't exceed her strengths. A small upper house was rented on the Eemskanaal on the edge of town, and, though the servants were sent away, a chairwoman was hired. Mother cooked herself and worked hard. The Groningeners didn't see the necessity of it all, but enjoyed her lust for work and the financial help that my father sometimes could offer. Did they care about judging others? In everything they were ahead of their times, for today what is in the common good was then forbidden. But their life was always full and happy.

For five years they lived in this small house, very happy with their simplicity and blue-collar life. My mother hated doing the dishes, a necessary evil, however; if only man could discover a magical way to make it a pleasure. Every evening she would deal with the problem, trying to get it done faster and more pleasantly. Great spirits eventually find themselves, for as father expressed it: "Fortunately, nature is put together such that every work, even the most boring, after awhile develops its own charm. The most boring machine – like calculating work (nothing in the world is more boring than machine-like number crunching) – has its appeal, as soon as man has developed it to its certain proficiency." Everyone did their boring machine-like work with happy enthusiasm.

After these five years they moved into a big old upper house on the Ossemarkt.

This was a house 'with soul'. People who visited it never forgot it. Although not fancy, the entrance was special: a small alley like a pin between two high houses and an unremarkable door with an old fashioned pull-bell. This, however, was seldom used since friends and acquaintances could open the door themselves so they could always come and go freely. The beautiful, old oak stairs led to a hall from which a big room opened with its air of serene rest and peacefulness, which benefitted everyone. The large family room was full of old-fashioned carpentry, antique drapes, and deep windows that were also high with the traditional Dutch half-drawn curtains that looked out upon the quiet square. Father's second laboratory, the former meteorological institute, given him in place of the reused commissioner's home with its friendly old style and tightly bricked entrance, could be seen on either side of the bridge. They lived happily in this house, which fit them so much and under whose spell everyone fell.

America

For the first time, father traveled to Mount Wilson in October 1908 on a three month government leave shortly after my brother, who had recently graduated and left for America to try his luck as a mining engineer. Father visited him in Denver on his way to and from California. Even in later years my parents were able to visit with my brother in America during their many trips to Pasadena, something few parents could do.

The first trip was not fulfilling for father. Without his wife he felt he was only half-a-person. She was the more talkative and easy going, who at once won accommodation with her carefree self confidence and open heartedness. Since his earliest years he had always admired how others came across with ease; as an adult he tended to be somewhat shy and timid. Very typical of his character, in 1913 he wrote me about it when again he was travelling alone:

> On board ship how does one go about getting to know anyone on board? I don't have a knack for it. Mother is much better at doing that and has broken the ice many times. Now that I am alone I feel my weakness twice as badly. In the meantime I have just had a lesson in doing that. While I was taking a walk on the deck, there came a man who had already struck me through his especially favorable appearances. As soon as he saw me, he came up to me and said, "Mr. Kapteyn, may I introduce myself, my name is Biesterman," or something like that, a grain man from Rotterdam. You see, as simple as hello. I think that I am too scared of a response. It won't do to be too sensitive in life. We book-learned people are in danger of that. Those who are among people all the time, on the exchange, in the office, while traveling, they lose that over sensitivity and therefore acquire readiness and ease, which I always envy in them. Although it seems to me that to bring the bow far into the art, men like van Dijk, say, still need to stand at the windows of his soul. He that lives too much inside himself, who keeps himself busy with things all the time, who

Fig. 11. J. C. Kapteyn, Karl Schwarzschild, and V. M. Slipher at the Mount Wilson Observatory, 1910. (Courtesy of Yerkes Observatory.)

cannot be of the daily life, seems to me never to be at home in ordinary living as others. Now I would not want to miss this inner life, but still instinctively my heart aches for that other life. Later when we just go around the whole world, daily to see other people, daily to have to adjust to new surroundings, then we will be real worldly people, on our 70th!

The basic cause, of course, went deeper than that and he understood that too. Man does not become a worldly man because of outside circumstances. Anyway mother possessed an easiness, though she still had no other chances in her calm, simple life. Now they complement each other nicely; together they lived life in all its fullness. Because she removed social difficulties that faced him, by making the first move towards making friends he made good friends everywhere. Additional steps others would make, those who got to know him and forthwith respected him. Thus, these trips were full of interesting discoveries and lasting friendships.

To his disappointment, during the first year of his stay on Mount Wilson the big instrument[30] wasn't ready. And Hale was too filled with undertaking his own researches on the Sun to be able to spend time on the stars. "So for the moment I don't have much to do. I do my own work that I would also do at home. Not very satisfying but it will get better." For resting, it was great on the mountain. It was sunny every day in November – the chilliest and most unfriendly month in the Dutch year. Yet he enjoyed nature's beauty, with a heart that was always

---

[30] Although the 60-inch telescope became operational in December of 1907, there were still various technical problems related to its use.

open to nature. The 2000-meter high mountain lay as a huge castle high above the Pacific, the clouds at its feet like an ocean of mist. As father expressed it: "The evening view over the valley, towards the Pacific, is often idealistically beautiful. All the pulls of the wrinkled valley soften themselves indescribably in the evening mist, and the colors of the evening sky above are in beautiful harmony with that."

He enjoyed life much more there when, the next year and for many years thereafter, mother accompanied him to America. On their many trips between Holland and America, they frequently visited Gill in London, who was always organizing gatherings of astronomers, through which father was able to remain in constant contact with this friend and English colleagues. Of these get togethers, Arthur S. Eddington wrote: "We rejoiced to hear again the familiar guttural exclamations and quaint expressions, as with youthful spirit and enthusiasm he unfolded his latest ideas."

Everywhere on their trips through America they were welcomed guests. They visited many observatories, such as those at Harvard, Yale, Princeton, Albany, and Allegheny, making their trips a real triumph. For father his yearly get togethers with American astronomers was of great importance. America had long since become a power in astronomical studies, not only through the spread of its resources, its enormous man-power (in the sense of jobs), and the energy of its astronomers, but also through her open-mindedness in dealing with the problems that were free from the pressures of tradition. This young nation had developed an energy and fresh strength that Europe knew nothing about, and to which father, young and enthusiastic, felt he really belonged.

They always stayed a few of days with Edward Pickering at the Harvard College Observatory in Cambridge, Massachusetts. Father admired Pickering's great strength and his ability to bring together a vast amount of material, which, although of a precise exactness, through his large staff allowed him to obtain a great reputation in the astronomical world. In 1912, Pickering proposed to father a cooperative project to work together on a northern photographic Durchmusterung. To make a catalog of the entire northern sky from declination 20° to the north pole of all stars to about the fourteenth magnitude. Father really did not want to undertake the project, but Pickering insisted. While Pickering would get credit for three-quarters of the work, he would also cover all the costs. After careful consideration, father decided to help Pickering. He knew that the work would be eminently useful, and with the thrust and the resources that Pickering had access to, this project had the potential to be finished quickly.

On the other hand, Hale was not thrilled with father's decision:

The value of the work is of course obvious enough, but a man of your ability ought not to be compelled to devote time and attention to such a piece of routine [as measuring and data reduction]. The more opportunity you have for thought on the larger phases of astronomical work, the more will astronomy benefit through the extraordinary range of your imaginative power. ... It seems such a pity that you should be merely collecting material for somebody else, incalculably less able than yourself.[31]

Father replied:

There is a sort of fate which makes me do all my life long just what I want to do least of all. The making of Durchmusterungen has no attraction whatever for me, but in what I have tried to do in the direction that has a true attraction for me, I have always been hindered by want of suitable material. So if – after I hope not too small a number of years – I come to die, I will probably leave behind me more Durchmusterung work and bringing together of material than almost anybody – leaving it to the next generation to do the real work that I hoped and longed to do. Well, when one doesn't have what one likes, it is necessary to like what one has.[32]

Besides the profit that the work would have, he hoped that this work would benefit science. For this big project he needed more employees, and he hoped, after completion of the astronomical reduction work, that they would be sent by the government to work in his laboratory at Groningen. "By that time I will be too old to profit very much by such an extension, but of course I have the future success of the laboratory much at heart." The cooperation with Pickering proved to be really very difficult. After the first weeks of organization, however, father no longer had to spend time on the work that his apprentices could do. Still, Pickering was not what Gill and Hale had been, and after a year father was sorry he had originally decided to take on this task. Because of other 'astronomical matters', the difficulties eventually became so great that they had to give up the project permanently.

Because my parents went to America every year, they had to relinquish their summer house, which put an end to their wonderful summers on the Drentsche moor. The family said their last good-byes with a nagging feeling of leaving something beautiful behind. The mayor and his wife came to say goodbye, and hoped to see them again sometime: "My wife and I have discussed that we don't know anyone that makes such good use of a decent income and can get so much happiness out of it as you." This made my parents very happy, because they had often discussed their life and their struggles here. Father really did understand the 'art of living', however. Although he never understood how to expand his material comforts, he knew how to use it for things that were extremely important and that were more fruitful than simply acquiring more possessions.

During their first summer at Mount Wilson they lived in a tent, because the observatory did not offer housing for married couples. Consequently, the mountain

---

[31] Hale to Kapteyn, 3 December 1912 (Hale).

[32] Kapteyn to Hale, 31 December 1912 (Hale). For Gill's concern, see Gill to Kapteyn, 27 March 1907, and Gill to Pickering, 25 October 1912 (K.A.L.), in which Gill chides Pickering for soliciting Kapteyn's assistance. In correspondence with Hale as early as 1905, Kapteyn expressed his devotion to the theoretical side of these questions, but lamented the fact of the paucity of relevant data; see Kapteyn to Hale, 7 May 1905 (Hale).

Fig. 12.   Kapteyn Cottage, ca. 1910.

came to be known as the 'monastery'. There were only small rooms available for
the observing astronomers, which they took turns using, while those who didn't
observe worked at the solar office in Pasadena. Thus their families lived in the
city. This resulted in a confined living that is usually unpleasant. Still, it seemed to
work very well because of the peaceful atmosphere, which for serious concentrated
work is preferred. For my parents, however, it was not very practical. The tent
was narrow and had few comforts. That is why they spent their days outside under
the open sky, which the warm reliable climate allowed. Tables, chairs, books,
everything were placed outside. Even though there were little bothersome flies,
that's where they worked. My parents, who moved from a pleasant, upperhouse in
the city to this pure natural living, still knew how to adjust themselves and to be
happy.

    The next year (1909) a big surprise awaited them. Arriving there they found a
small wooden house that Hale had decorated with every possible comfort. Everyone on Mount Wilson had contributed something to it. The staff continually asked,
"Now if it were to be yours, how would you like this or that to be done?" And
with all their help, in all its simplicity it became a little jewel. The 'Kapteyn
cottage', as it is still presently called, became their American home; so for many
years it became equally as precious to them as their Dutch home. It was built on
a wonderful spot where shadows were cast upon it by knotted oaks and very old
evergreens and where large white yuccas stood as giant bouquets around the house.
The view was grand, far over the mountains and the canyons the deep sleeping
ravines of western America with its valleys and many states, in the evenings like
lighted bunches of stars the feeling of rest was heightened high above it. At the

same time, however, it reflected a lonely feeling that human life was nearby. And in the distance lay the majestic waters of the Pacific Ocean. The peaceful quiet that reigned far from the hunting world was what they yearned for doubly after an exhausting and troubled voyage. Deer came curiously to look around, squirrels scurried silently from trunk to trunk, and the chickadees, the titmice of the West, cackled the whole day long. The cool nights were wonderful and mother set herself up on the big veranda that stretched the length of the house for the evening. One night she saw something dark on her bed. It seemed to be a nest of young squirrels, for whom the mother had chosen this spot to bring them into the world. Idyllic, but still a little creepy thought mother. And so she looked for somewhere else to sleep that night.

My father missed heather that Dutch plant he loved so much, and so he decided to plant them. The next year he traveled with a box of heather as if it were costly and fragile baggage that required all his attention. Once, when it remained behind in the train as a result of which it was unexpectedly rearranged, he called and telegraphed all over until it was finally returned to him. Eventually, it arrived undamaged at the mountain and was planted by the house. Regardless of all the love and care, the heather could not set its roots. Thus he had to give up his wish.

Those who came to visit father were heartily welcomed. Father always had advice and an interesting word for everyone. And mother quickly won other's trust, and so it became a rich life for them.

The mountain air was an elixir of life that gave him double the strength and desire to work. He had no constitution for the climate of the valleys that made him heavy and tired, and where he continuously sought the use of quinine. In the mountain air, however, he felt like another person with greater vitality and unlimited power. "I feel youthful cravings awakening," he wrote, "to cling around in the mountains, descend into the canyons, and because I can not give into that, I have recently tried my strength at difficult trips along steep and badly kept mountain trails."

With the Americans he felt at home. The unconventionality and simplicity of my parents fit completely into the life-style of these open-minded and uncritical people of the Far West, who did not resonate with the centuries old culture with its requirements, conventions, and sentimentalities. It is with this attitude that my parents were greeted with the most heartfelt welcome each time. "We are glad to see you folks again," they all said with happiness in their eyes and warmth in their hand shake.

In 1913, father wrote to his friend Boissevain: "I have had a wonderful, busy time in America. Parties, parties, and parties. Speeches and trumpet sounds. Vanities vanitatum. Oh, yes, but the American does it in a more cheerfully and childishly way than we do. And I found much thankfulness." It was no wonder, because everyone was affected by his personality. After a visit at Yerkes, the astronomer E. E. Barnard[33] wrote:

> I simply wish to tell you, how much we enjoyed the short stay you made at the
> Yerkes Observatory. It is a great pleasure to see you here, and it always leaves
> behind a recollection that makes me feel good for a long time afterwards.[34]

Whenever he came, all the observers would ask his advice and guidance in helping
them set-up their work programs, in judging how to use their instruments, and
eventually in helping them choose their strengths.

He enjoyed the energy and efficiency of the Americans, who to Europeans
bordered on the incredible. For instance, if he had an idea early in the morning,
he would send his list to the solar office in Pasadena with a statement about the
stars, indicating he needed calculations, notes, and conclusions. The next day they
would return his notes to him complete; this method of work was to his liking.

Like a temple of knowledge, the Mount Wilson Observatory lay there high above
the clouds. With religious seriousness and devotion its 'servants' worked. Other
than religion, what is knowledge but the search for truth, for the eternal power
that guides the world and keeps it in her tracks. Above all, America supports
knowledge, both financially and socially. When energy and devotion are united
anything is possible with such enormous spiritual and material resources.

The stay on Mount Wilson, however, was not always appealing to father. As he
wrote in 1913:

> I don't believe I can handle the California trips for many more years. It would
> be the most beautiful, nicest position thinkable – if it were not paid. But since
> that is not the case, I sometimes wonder if I am giving my money's worth. I
> would like to answer that with a 'yes', if I could simply put the men here to
> work to do, what in my eyes is necessary. But of course I only have a small
> voice in giving advice, a voice that is listened to, but still ... the whole set-up is
> for 'solar research', for astrophysics, and I represent the other side [i.e., stellar
> research]. The officials are all on the other side. And yet they have done very,
> very much to my liking with me in mind. That is the sickness of doubtfulness
> that is deep in my blood.

What was this 'sickness of doubt', because wasn't his presence and inspiration,
his example and stimulation, his enlightenment and guidance, not payable enough?
On the fortieth anniversary of his receiving his professorship, Hale wrote father of
his own good luck:

> You have given a marvelous illustration of the possibilities of work by an obser-
> vatory without a telescope. You have also stimulated astronomical research
> throughout the world in an extraordinary way and initiated an undertaking

---

[33] Edward Emerson Barnard (1857–1923) was the foremost observational astronomer of his time.
He was observer at Lick Observatory in California (1888–1895); in 1895 he was appointed professor
of astronomy at the newly founded University of Chicago and observer at Yerkes Observatory where
he remained until his death. He is best known for his discoveries of comets, nebulae, and particularly
Jupiter's fifth satellite. He was the first astronomer to photograph the Milky Way.

[34] E. E. Barnard to J. C. Kapteyn, September 1913 (Yerkes).

which will be continued far into the future. Under such circumstances you must surely find pleasure in reviewing the years of your worldwide work of cooperation, which has meant so much to astronomers everywhere. Most of all the Mt. Wilson Observatory is deeply indebted to you for the greatly broadened conception of its possibilities, which you have awakened. Every member of the staff would thus join with me in the assurance of their cordial appreciation of the aid and inspiration which we owe to you.[35]

Being a modest person, father did not promote his own achievements, and thus he always considered himself as the one who received most. Occasionally he felt tired from the pressure of all the problems that concerned him that left him little time for other things for which his soul longed. From Mount Wilson he wrote me: "My life is nearly consumed with problems of knowledge – and seldom do I bring them to some kind of satisfactory resolution, and then it's off again to search for something better. Then sometimes I think like Cauchy: 'Oh what sad employment/ What humiliating feebleness/ To calculate always/ To integrate without stopping.' But it stays with the old, and I shall stay with it even if sometimes I have a yearning for my retirement so that I can put the arithmetic marker down and enjoy all that is human. Will I still be able to do that?"[36] But this was only a temporary melancholy. Quickly he shed himself of this feeling so that with renewed strength and enthusiasm he could attack the problems without which he could not live. And he found in this happy optimism a fresh buoyancy, so that in subsequent letters there was a lot more vitality.

Mother did her best to make their stay on the mountain homey and cozy. Even though the housework was really not that easy, she did all of it herself. Everything had to be ordered by phone from Pasadena, and too often it happened that the wrong things were brought, which then had to be returned and exchanged. Thus everything had to be handled from a distance of twelve kilometers. But she enjoyed overcoming these difficulties and not getting discouraged.

During her first days in Pasadena she would have to travel to Los Angeles, the largest commercial city bordering Pasadena, to do the shopping. The stores offered a variety of household articles and many elegant new products that are very attractive to the good housewife and mother.

When our parents returned to Groningen they always brought a suitcase full of surprises – delicious American goodies that were ingenious, practical, and graceful. Together they had thought of various monumental presents and bought them on their return, such as a typical American swinging chair, a huge lawn swing camouflaged as a flying machine that gave them much trouble in transport for the grandchildren when inspected by customs in Liverpool. But nothing was too much

---

[35] G. E. Hale to J. C. Kapteyn, 22 January 1918 (Hale).

[36] A mathematician of enormous ability, the Frenchman Augustin-Louis Cauchy (1789–1857) worked on the calculus placing analysis on a rigorous foundation, made fundamental contributions to complex function theory, contributed significantly to error theory, algebra, geometry, and differential equations, and made numerous contributions to celestial mechanics.

for them if they could bring happiness through it.

Father always returned from America full of new plans and ideas, renewed with health and strength, and longing for the laboratory with its quiet life that he needed for concentrated work. So it was each time – lucky to leave and lucky to return. Both my mother and father appreciated this exceptional, happy life as much as they could.

The Great War

At the beginning of 1914, father encountered a heavy blow. After a short though not severe illness, Gill died. Deeply saddened, my parents went to London to see their loyal friend for the last time. They found Lady Gill in deep mourning. She lay exhausted in bed, waiting for my father as her husband's most trusted friend. He knelt by her bedside, and putting her hand solemnly on his head, she blessed and thanked him for what he meant to her husband. All the love and gratitude, all the grief and mourning of a great human heart were expressed in this impressive gesture.

Gill's wife and some of his friends accompanied him to Aberdeen, his place of birth where he was to be buried. At the same time, a memorial service was held at St. Mary Abbot at Kensington in London where my parents and many sad friends had gathered. The organ played the beautiful song from Lord Tennyson: 'Crossing the Bar', and a pure boy's voice like heavenly music sang the comforting words. They returned to Gill's home quietly, packed their suitcase and left '34 de Vere Gardens', that hospitable house where they would never enter again, as lady Gill would soon leave it permanently. They returned home poorer. Every Sunday morning mother played this melancholy death song, and we heard in the quiet morning hours, its tunes rustling through our home. And we were quiet in memory of a loyal friend.

This was the sad beginning for the disastrous year of 1914 that would press its ominous seal upon the world leaving death and destruction. Even knowledge – the untouchable – that one had thought enthroned above all the world's happenings would have to sacrifice to the god of war, which for father was a great deception.

At the outbreak of the Great War, father and mother were at Mount Wilson. As if with one blow, everything had changed all over the world. Everyone lived in fear and worry about distant relatives and friends. Those who were able to go home did so as soon as possible. As for my parents, however, a trip crossing the ocean with the danger of mines was impossible, so they stayed until January, and then, filled with a tense fear, came safely home. It was their last trip on the ocean – they would never see America again.

In July of 1914, shortly before the outbreak of the war, father was awarded the Ordre pour le Mérite by the German Kaiser, something that he accepted with great joy. "I don't understand, why this is happening to me," he later told his assistant Pieter van Rhijn. This award, which was only adjudged to thirty foreigners, was considered one of the highest distinctions in the scientific world. When Gill received this award in 1892, he noted: "The Ordre for the Mérite I regard as the highest distinction open to a literary or scientific man." At the beginning of August, however, when there was news in America that Germany had violated Dutch neutrality, father felt obligated to refuse the award. On 6 August 1914, my father wrote the German Consul in Groningen: "Because Dutch neutrality has been violated by the German army, I feel obligated to reconsider and change my decision of July 28 (of this year). Under the present circumstances I cannot accept the appointment as foreign member of the Ordre pour le Mérite, which was awarded to me by the Emperor of Germany and King of Prussia. I sincerely hope that this letter will get to you on time." The Consul wrote father back telling him that the news was not true, which consequently negated father's reason for refusing the award.

In England, however, people were indignant that my father was accepting a German award. Some hotheads among the astronomers reproached him, because they, as so many during that time, only saw their own point of view, notwithstanding their scientific education.[37] Father could understand this neither now nor later. Perhaps it is impossible for a neutral to understand the mentality of war-making nations with the suffering of body and soul, the fear and uncertainty, and the incredible tensions. Such things resulted in a near insane bitterness and hate, making any form of objectivity impossible. Even a scientific person was a human above all else. Who, however, could remain objective? As the eminent German astronomer Karl Schwarzschild expressed very humanly: "If everyone is wild, why shouldn't I be wild, too?" One became part of the battling, suffering, and hating community, nothing more and nothing less.

As Hubrecht, son of one of my father's friends who was an astronomer at Cambridge, wrote at the time: "May I offer you my sincerest congratulations upon the opportunity of the unique distinction awarded to you by the German Emperor. My brother wrote me about it. Otherwise, the news has not made the English papers yet. I think the consensus will find that the Emperor will be viewed from a more favorable point of view through all of this. The few English friends I have spoken to are at least surprised and could really not have imagined him doing such a culturally confessing deed."[38]

The horrible war years brought about a division in the scientific world, which, after termination of the hostilities, healed very slowly. Father wanted to help bring about unity, but his idealism did not count on the cruel reality, and for the first time in his life, his tact and strength were not sufficient. As it was impossible for him to travel to America during the war years, he felt a burden on the Carnegie Institution and thus he wrote Hale that he wanted to withdraw as Research Associate. Robert

---

[37] See G. E. Hale to Willem de Sitter, 10 February 1921, W. de Sitter to G. E. Hale, 9 March 1921, and G. E. Hale to Frank W. Dyson, 30 March 1921 (Hale).
[38] Hubrecht to J. C. Kapteyn, 27 August 1914.

S. Woodward[39], President of the Carnegie Institution, persuaded him to stay with the help of Hale, because father's work was of great value to the Institution. It was inconsequential whether father visited America regularly or whether he stayed home. Notwithstanding the transportation problems, father wrote his friends in America regularly and was thus able to continue his working arrangement with Hale. The latter, however, became more and more influenced by the war, and after the destruction of the Lusitania (by torpedoes), Hale wrote in 1915: "I have lost all patience with the Germans, since they turned pirates, and would not blame England for the most drastic action, whether it injures us or not." He offered his services to the American government for 'Preparedness for war'. And until the end of the war, this remained his major line of work.

Father saw the increasing hatred of the allies and America toward Germany. His feelings of justice took up against the abuse of the German people, who were not guiltier than any others. He saw clearly through the centuries how history kept repeating itself, and how every country, at its turn, had done wrong whenever it had the power to. Wasn't the Boer War still fresh in everyone's memory, that dark page in England's history? Had France gone freely into Morocco and the Dutch into Indonesia? And was the barricade, that had brought Germany to the verge of starvation, any more humane? A hard-working nation with many virtues had become blinded by its sudden revival. It became poisoned by a system of military slogans and patriotic phrases and was being ruined by its slavish belief in a government that was not equal to its task. Thus he saw the tragedy of this nation's people; he could do nothing but take up for the repressed, who were no more barbarian than any others.

By the end of 1918 hope had arrived. Woodrow Wilson, President of the United States, whom everyone knew about his peace-loving and constructive ideals, as the prophet of the new regime would come to Europe and help formulate conditions for peace. He was compared to the profit of a more beautiful cooperation, and many people shared my father's feelings of a hopeful expectation by seeing him.

As father wrote the astronomer Edwin B. Frost[40] at Yerkes in 1918: "But for your admirable president I for one feel but little hope of an issue which to humanity would be worth the awful misery it has gone through. Again, I for one have no fear of Wilson. After Christ he may become the greatest benefactor of the world. I feel sure that he will fight for his ideals or perish in the attempt. Better forego rightful vengeance, rightful punishment, anything than the hope for some enduring settlement of human affairs that will make for the real happiness and progress of mankind. There never was such a chance. There would not be now, but for

---

[39] The American R. S. Woodward (1849–1924) was trained as a geologist and became president of the Carnegie Institution of Washington in 1904 serving until 1920.

[40] The American E. B. Frost (1866–1935) succeeded George Ellery Hale as director of Yerkes Observatory in 1905 serving until 1932. He became a specialist in stellar spectroscopy, specifically the determination of the radial velocities of stars.

America."[41]

Unfortunately, the appearance and as quickly the disappearance of President Wilson became a tragedy. This weak man, the idealistic professor who knew nothing of European relations in faraway America, had to compete with the cunning diplomats of the old world. "Even though strength may fail, will-power is to be praised." Because of this will, however much she failed in Versailles, the foundations for a slowly developing peace and brotherly beginning was laid in the League of Nations.

Instead of being conciliatory, however, the peace at Versailles put irreconciliation in hand, and long after the enduring effects of the war, it glowed and boiled everywhere both in the hearts of the defeated as well as in the hearts of the victors. Science, too, kept choosing sides, and in July of 1919, the 'International Research Council' was established – excluding Germany.

With Heymans my father discussed what they could do to bring about a reconciliation within the scientific world. He had an unwavering trust in science and its ultimate blessing over this bitter subjectivity. As he wrote Eddington in 1917: "It is my conviction that science must in the long run directly and indirectly become a mighty factor in bringing peace and goodwill among men. If the men of science give an example of hate and narrow mindedness, who is going to lead the way?"[42] In this great English astronomer, a Quaker and a strict scientific man, he found a human and an unbiased judge who did not let himself be blinded by hate. But Eddington was an exception at that time.

After much discussion, father and Heymans composed an open letter in 1919, entitled 'To the Members of the Academies of the Allied Nations and of the United States of America', in which they swore that the sciences were the "great conciliator and benefactor of mankind." Closing they wrote:

> We understand how your attention of late has been monopolized by what is temporal and transitory. But now you more than all others are called upon to find again the way to what is eternal. You possess the inclination for objective thought, the wide range of vision, the discretion, the habit of self-criticism. Of you we had expected the first step for the restoration of lacerated Europe. We call on you for cooperation in order to prevent science from becoming divided, for the first time and for an indefinite period, into hostile political camps.[43]

Writing the mathematician Diederik Korteweg[44] in 14 August 1919, father expressed:

> I do not imagine that this open letter to the allied academy members will undo

---

[41] J. C. Kapteyn to E. B. Frost, 10 November 1918 (Yerkes).

[42] J. C. Kapteyn to A. S. Eddington, January 1917.

[43] Draft of letter, William W. Campbell to *Science* magazine, December 1919, (Hale Mss, Box 61, NRC Division of Foreign Relations file). For a discussion of the circumstances involved in this important episode, see Daniel J. Kevles, "'Into Hostile Political Camps': The Reorganization of International Science in World War I," *Isis*, 62 (1971), particularly 59–60 (47–60).

[44] The Dutchman D. J. Korteweg (1848–1941) was professor of mathematics at the University of Amsterdam from 1881 until his retirement in 1918.

the tearing and association of the academics. I have not abandoned all hope, however, that such a letter, signed by a great number of highly-positioned neutrals, will bring the allies to a more objective view of things, with the consequence that there will be less haste in forming interallied trade unions. In my opinion, when time is won, all is won. If this does not work either, then I still shall not regret the decision I took with Heymans.[45]

The time was not right, however, for their plea amount to nothing. On the contrary, the interference of neutrals, who had completely stood outside of the horrible battle with all its misery and desperation, irritated rather than stimulated thinking. Feelings still ran high in opposition because the wounds were still so fresh.

At the end of 1919, when a summons was sent out to the neutral academies to affiliate with the International Research Council, excluding Germany, father resisted with all his power, and tried to keep the Amsterdam Academy of Sciences from taking this step. In the decisive meeting, he and Heymans used all their influence to keep the members from agreeing, but they were unsuccessful. Here was the opportunism against idealism, and, in science too, opportunism won out. He had not expected it and the shock was so great that Heymans immediately resigned from the Academy and father never again attended Academy sessions. By its own standards, it had proven its inability to handle situations in a just and scientific way. It was more than a passing shock, though. Until his death in 1922, this complete failure of trust in justice and objectivity remained a painful wound.

S. E. Strömgren[46], director of the observatory in Copenhagen, also sought for reconciliation. Like most Swedes who had an unlimited admiration for German science, Strömgren, whose personal sympathies were for Germany, sided naturally with the Germans. He was upset about the exclusion of the Germans, and when, in 1919, the French refused to let data on the comet circulate to Germany – something that Harvard allowed – Strömgren became indignant: "I have two boys aged 9 and 11, who wouldn't do a thing like that." Through mediation he was able to reach many Germans during those difficult times of isolation. In 1920 when they decided for the first time in a long while to hold meetings of the Astronomische Gesellschaft, he worked, as president-elect, with all his power for the meetings to be successful. He suggested that my father become a member of the steering committee of the Gesellschaft:

The decision, if you understand it, will bring you temporary discord with Baillaud, Lecointe, and Turner.[47] But for this the entire neutral world will be

---

[45] The Kapteyn-Heymans letter eventually contained 278 signatories drawn from the neutral countries Denmark, Sweden, Holland, Norway, Finland, Switzerland, and Spain.

[46] The Swedish astronomer S. E. Strömgren (1870–1947) was professor of astronomy and director of the observatory at Copenhagen University from 1907 until 1940; he was a specialist in classical celestial mechanics.

[47] The French scientists Jules Baillaud and M. Lecointe and the English astronomer H. H. Turner were among the most militant in arguing for exclusion of German scientists after the war.

thankful to you, and a large number of the allied astronomers would greet this decision with sympathy – of the feelings of German science I do not need to speak. ... You have given science so much – I believe you could again bring a lasting usefulness.

My father decided to accept. As he wrote Hugo von Seeliger[48], the dean of German astronomy in Munich:

I am fully aware that generally I do not have the qualities really needed to be a good committee member. That is the reason I have always used to turn down such offers. Now I believe that the situation is somewhat different. It is now necessary that I offer my opinion in order to protest this irresponsible failure so that those with the best abilities may assist in helping to heal this destructive rupture.

Eddington was the first among of the hostile camp who was ready for reconciliation. Writing Strömgren in November of 1919: "I hope to show my interest in the Astronomische Gesellschaft by attending the next meeting – an individual step which no one has any right to object to. ... International science is bound to win and recent events – the verification of Einstein's theory – has made a tremendous difference in the last month."

In 1921, Eddington became the only Englishman to attend the meetings of the Astronomische Gesellschaft held in Potsdam, Germany. As we write in 1928 now, Germany has been fully accepted into the League of Nations. The ideas of Wilson, who at the time was the greatest doer of good and reconciliation, is helping heal the wounds and tears in science.

---

[48] Along with Kapteyn, Hugo R. von Seeliger (1849–1924) was most responsible for developing statistical astronomy into a major research branch of astronomy. He spent his most productive years as professor of astronomy and director of the observatory at the University of Munich (1882–1924); he was president of the Astronomische Gesellschaft for 24 years (1897–1921). See Paul, Robert E.: 1993, *The Milky Way Galaxy and Statistical Cosmology*, chapters 3 and 6.

CHAPTER 6

LAST YEARS

Leiden Once Again

After they had tried in vain to appoint my father to the directorship of the Leiden Observatory, van de Sande Bakhuyzen was succeeded by his brother Ernst in 1897. And when Ernst Bakhuyzen died in 1918, Willem de Sitter became director of the observatory. De Sitter pondered a complete reorganization in order to bring the observatory to greater efficiency. Much unfinished material had piled up and he wished to put things onto a new track. After consulting with my father, he decided on an entirely new direction. Together with himself, there would be two adjunct-directors named who would tackle the three goals he had outlined. First, as director de Sitter himself would direct the department of theoretical astronomy. Second, Anton Pannekoek would direct the department of positional astronomy, because he had done this work at the observatory and was acquainted with the technical requirements. And third, Ejnar Hertzsprung[1], who was my husband, would be responsible for the newly created astrophysics department. This arrangement seemed destined to great success. Although Hertzsprung was named, for political reasons the Ministry was set against naming Pannekoek as a government official.[2] The latter opening remained unfilled since a suitable replacement could not be found in either Holland or elsewhere. Both de Sitter and Hertzsprung set to work with all their energy, but they sorely missed their third man because the positional work, which was pressing to be completed, remained undone.

When my father was visiting us in Leiden, he discussed the problem with my husband, Ejnar Hertzsprung. Suddenly I had a thought and blurted out: "Father, couldn't you temporarily fill in? You've had a handle on the work for years. And because of your emeritus status, you'll have a lot more free time!" Hertzsprung supported this impulsive proposal. But father laughed about it for this idea hadn't occurred to him. After some discussion he said, "It isn't such a crazy idea really and I shall give it some serious consideration." He concluded that the situation needed immediate attention to bring order to the present chaos, and felt that he, more than anyone, was the best choice for the job.

---

[1] The Dane E. Hertzsprung (1873–1967) was astronomer at the Astrophysical Observatory at Potsdam (1909–1919) joining the observatory at Leiden in 1919 becoming its director in 1935 until his retirement in 1944. Known for his studies in stellar evolution which, along with the work of Henry Norris Russell (1877–1957), have become embodied in the Hertzsprung-Russell diagram.

[2] Pannekoek was a militant supporter of Marxism throughout his entire life; see Bart Bok interview, pp. 14–15.

The Leiden Observatory had always held his love. He felt home there and was fully aware of its problems and situation, and he knew the staff and was appreciated by all. It all seemed right. In April 1920 he wrote Ejnar:

Now that you take it seriously, I've also thought more about it. I will accept a position as 'advisor', and work just one day per week.... This is inspired solely through the detailed plan by yourself and only to help the Observatory. Personally, this proposition doesn't have anything new to attract me....[3]

This cost him a bit of his freedom of movement, which he so longed for. "It is precisely this complete freedom, which for me makes this emeritus status so attractive," he wrote to Gustav Eberhard[4] at Potsdam with whom he had become good friends over the last years.

The clincher was that de Sitter was totally pleased with the proposition that father would temporarily become an adjunct-director in which he would spend one week out of two months in Leiden. He would stay with his children, which was a great joy for all involved. "Now you'll be an assistant to your former assistant," they teased, but they were also proud of the greatness of his spirit that he would attach such importance to this problem. There were those who wanted to give him a special title, because they deemed this function beneath his position and too humble a job for him. But such social considerations were foreign to him and of no interest, whereas love lifted him higher than title could ever have done. So he came back to the Leiden Observatory, where almost a half-century earlier he had begun his career. He planned new observations that primarily dealt with testing the fundamental stellar positions, as well as supervising the results of the older Leiden observers.

It was only a short stay, for within a year the incurable sickness that put an end to his works appeared. But in that one year, he organized and provided order to everything and put it all on the right track so that others could continue the work.

Last Years at Groningen

In 1918 my parents had to leave their beloved house on the Ossemarkt because the landlord wanted to live in it himself. It was a time of housing shortage that prevailed everywhere following the war. So they decided to spend the last two years before his emeritus status in a hotel, as they would later spend their retirement in Hilversum. Consequently, they moved to the old renowned Hotel de Doelen on the Groote Markt where two rooms, filled with their own furniture, made for a comfortable home. One of them was a big family room at the front of the house with plenty of space for the grand piano and an old-fashioned canopy. This room looked out over the big market square with the stately Martini-tower at the right

---

[3] Kapteyn to E. Hertzsprung, 18 April 1920.
[4] P. A. J. G. Eberhard (1867–1940) received his doctorate in astronomy under Hugo von Seeliger at Munich, and from 1916 until his retirement he was senior observer at the Astrophysical Observatory at Potsdam.

and the massive City Hall at the left. Tuesday and Fridays was market day, which brought with it a pleasant commotion. You could sit for hours on the big window sill and look out at the busy square and never get tired of it. Every beautiful May, however, brought with it a noise that blared and was tumultuous but which put the carnival right at their door with all of its merry-go-rounds and hippodromes that churned out whistles and played music and filled the spring air with smells of buttered and sugared tiny pancakes (called poffertjes) and hot machine-smells. Groningen celebrated a festival, because it didn't want to lose the opportunity. That was a busy, but unpleasant time for the guests at the hotel. Often my parents took off and sought rest elsewhere, because they couldn't find any at home.

At the Martini-churchyard, hardly a five minute walk from their hotel, lived a young professor named Bordewijk, who did not feel at home in that strange Groningen, which forty years ago mother had also struggled with. Because of their spontaneity, sincerity, and same-mindedness, he and his impulsive wife found their way quickly into the hearts of my parents. They were welcomed with so much warmth that a deep friendship developed between them. In later years, Mrs. Bordewijk was asked how it was that they had developed such an intimate relationship with them since they were so much younger. "I don't know how it came," she answered, "but I do know that they were the most beautiful years for us in Groningen."

For father this contact with the young was a real joy. His spirit had always remained young. It could withstand everything and went along with it all, enjoying the enthusiasm and the elasticity of youth that the old often lose.

Happily and without hesitation father could go the distance that a younger generation had gone. Especially as a learned man this was so. Youth drew strength from his experience and his optimism, and breathed in the serene rest that lay over his home life. This simple but great man taught them by his example to see the truly great. Wasn't life glorious and worthy of struggle? All problems and trifles could be solved in pure harmony. They felt new strength and courage in themselves to realize their part in that harmony. And with this blessing, they returned happy to their homes. Thus one of the dearest of all their youthful friends called them 'Father and Mother Kapteyn'. They had a big international heart for in America, England, and Germany there were friends who had this privilege, and with his fair name came also the beautiful trust that can exist between people.

Numerous knighthoods and decorations, medals, honorary doctorates, and so on were accorded father during the years of his scientific work. The first order, the Legion d'Honneur of France, had brought him much happiness. As the first official recognition of his work, father always wore the red ribbon in his button hole. Further distinctions that he also valued my father viewed primarily as a way of getting what he needed for his scientific work and also to secure a better salary for his clerks. The public fame and fanfare was appreciated because it gave him the means that served him and his cause to grow at an ever increasing pace. So it happened as well at that time. In 1903 the former physiological laboratory, which contained the same two rooms where as a young man he had begun his scientific work, was given to him because of the influence of his friend Huizinga. Here an assistant and several clerks went about their daily labors where great work was accomplished.

Far deeper than the worldly renown was given to him through the love of people. They stirred the deepest part of his soul and made him happier than any worldly success could. The most impressed I ever saw him was due to a letter from the astronomer Gustav Eberhard at Potsdam, with whom he had a deep friendship during his last years. During a dire sickness, Eberhard wrote my father how much their friendship meant and that it was the best friendship that his later years had brought him:

> I thank destiny especially that it has permitted me in the last two years to have come close to you. In any case, when a new publication of yours arrives, one studies it as if one were in a joyous mood.

For father this honor was also clearly expressed in an earlier letter from Robert Innes[5], who father had invited to call him by his first name:

> It is very kind of you to admit me into the inner circle by addressing me as 'my dear Innes' and more kind to give me permission to address you in the same way. If I do not do so, it is not because I do not value the permission, but because your truly eminent talents place you on a higher sphere. I do hope you will not object to my hero worship (which for my part I think a very good thing for a man to be capable of)....[6]

Love, for him the highest quality, he shared completely with others. He wanted to see people happy and help them with their needs. The poor sales women who came to his laboratory to offer their wares for sale, he could not turn away even though he knew her wares were poor in quality. The closet was full of rough writing paper and bad pencils for which he was always able to find a home. For he, the blessed, wanted to help those who knocked on his door. He was a bottomless well when it came to helping and counselling whenever his family, friends, clerks, and others needed him.

For example, his former servant Ottens, who had become an invalid from a fall and therefore could no longer work, father knew how to help. Father went around to all of his friends and colleagues and asked them to take their bikes regularly to this man to be cleaned. Father saw to it that Ottens could take care of it himself by getting him the necessary materials to do the job. The business became a success and in this way the poor Ottens could take care of his family. This is just one example of many.

---

[5] R. T. A. Innes (1861–1933) was director of the Transvaal Observatory in Johannesburg from 1903 until 1927. A talented observationalist, he is known as the discoverer of Proxima Centauri, the nearest known star to the solar system.

[6] R. T. A. Innes to J. C. Kapteyn, 14 April 1906.

It was moving to see my father with his grandchildren. His untiring interest and patience for them, and the plans for their future were endless. For some he started a bank account for school when they were born and he always knew how to counsel regarding their up-bringing. Even on his death bed, he was making plans: Each child would plant a tree in the garden of my parent's summer house where he would live out his retirement and thus watch them grow. How glorious it would be to sleep outside with him, learn about the birds, make wild-flower bouquets, and read with him! No wonder that the children were totally in love with their grandfather. Especially after his death can we see this. One of his grandchildren, who had just begun to learn some history at school, was repeating her lessons that dealt with the counting of the years. Her mother asked, "But Rigel, what does that 1500 mean?" The child did not know, because her teachers had yet to explain it. So her mother explained, "1500 years ago a man was born whose birth heralded the first year of the way we count the years. This man was the best, most noblest, and the wisest of all men that ever lived; can you guess who this man was?" "Grandfather!" shouted the child without a moments hesitation.

For the sake of science, father also wanted to take care of his students financially. Thus he established a fund, of which he himself laid the foundation. It would now be possible for those without means but full of talent to study astronomy and thus to engage in astronomical discoveries. But scientific people are not financial people; consequently, funds grew slowly. Once he met one of the friends of my older sister, W. Dekking, a tradesman from Rotterdam. Impressed with father's personality and struggles, Dekking recommended that father turn to the tradesworld in Rotterdam offering himself as the one to supervise the funds. Because of Dekking's enthusiasm and hard work, the funds made impressive growth. With the help of many friends and admirers, following father's death it became institutionalized under the name Kapteyn-Funds, an official institution of considerable means. Already it has been of great service to many and shall continue to do so.

Father's seventieth birthday was on 19 January 1921. It was not a day of joy and celebration, however, but a day of tenseness and quiet waiting. On this day father waited with his family in Amsterdam for my sister, her husband, and their child, who were returning from America. The famous brain surgeon Harvey Williams Cushing[7] had operated on my sister's child in Boston, and they were expected back that day. The future was uncertain; but via the wireless it was signaled that everything had gone well on the boat home and everyone breathed easier. Still, father felt beaten by this first real pain in his own family. The 'Kapteyn-luck' was a saying in Groningen: Prosperity was all this family had known and nothing had ever beaten them down until this dear little grandchild had suddenly shown the symptoms of a serious brain illness. Father supported the sorrowful parents with all his strength and love by sending helpful, comforting letters during their stay

---

[7] H. W. Cushing (1869–1939) became the first American to devote full-time to neurological surgery developing a world-wide reputation as professor of surgery while at Harvard. He was surgeon-in-chief at Peter Bent Brigham Hospital from 1913 until his retirement in 1932.

in America, and waited now to receive them again in love and help them carry their pain. It was a day of joy when they were happily reunited again, they who so loved each other and were so bound to one another. Of the outside world father did not notice much. Everyone knew that this day would have to be celebrated by father quietly; so they respected father's sorrow. The day that ushered in father's retirement, passed without fanfare.

Shortly thereafter father received notice from the government that announced the decision to put his name on the laboratory as a permanent tribute to its founder.

In June father gave his last lecture, and my parents left Groningen. Close friends were invited in turn to their table at the Dolen. My parents found this to be a more intimate way to gather friends together than at a huge dinner that offers more sparkle than intimacy. And when they said farewell to the hotel, it was with melancholy that they left everything behind. The manager said her good-byes tearfully and said that they had been the easiest guests she had ever had. Feelings were mutual, for they had been good years indeed. At 7 o'clock in the morning they left. None of their friends knew the time of their departure so that no one could hold them back. This was how they had wanted it, and quietly they said good-bye to the city where they had lived and worked for forty-three years. A long and happy life of fruitful labor, of a happy home and of good friendships now lay behind them, but a promising new life still lay ahead for this couple who, though in the twilight of their years, still possessed light and strength to meet new challenges. First, however, they wanted to enjoy a vacation in a clear climate where father could regain his strength after the stress of the last few years. The last years before his retirement, father had pushed himself to the limits of his strength. He had avoided everything that would have taken from his costly time: Responsibilities, meetings, even a visit to Mount Wilson that he wished to visit very much. "I have to give all my time, including my vacations to the laboratory," father wrote Hale in America. Everything must be put in order and completed so they could rest easy at his departure so that his former student and now assistant Pieter van Rhijn could continue the work as his replacement. At the same time, however, an important new theory based on the results of his work of the last few years so enthralled him and kept him so busy that his books and papers accompanied him on his vacation to Switzerland.

This theory was the first step on the way to an explanation of the dynamical structure of the stellar system based on his observations. It was not a complete, definitive theory, but more the result of the first discoveries of the unknown, beyond the boundaries of certain knowledge. My father himself named it, "A first attempt at the theory of the arrangement and motion of the sidereal system."[8] Father first presented his ideas to the 1921 Dutch Astronomers Club after his return from the Astronomische Gesellschaft meetings at Potsdam, where he with Albert

---

[8] Kapteyn's theory was published in 1922 as 'First Attempt at a Theory of the Arrangement and Motion of the Sidereal System', *The Astrophysical Journal*, 55 (1922), 302–28.

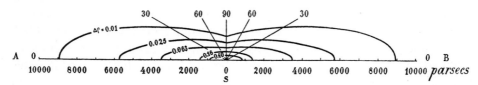

Fig. 13.   Kapteyn's 1920 Stellar System. The curves represent lines of equal density distribution perpendicular to the plane AB of the Galaxy. The numbers 0, 30, 60, and 90 represent galactic latitudes in degrees. Density numbers are relative, with the density of the Sun (assumed to be at the center) taken as unity. (Reprinted with permission from *The Astrophysical Journal*.)

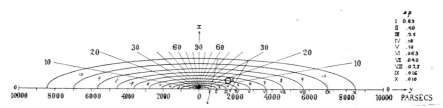

Fig. 14.   The 'Kapteyn Universe' (1922). The ellipsoids of revolution represent lines of equal density surfaces in a plane perpendicular to the plane of the Galaxy. Marginal numbers represent galactic latitudes; roman numerals represent (relative) densities with the center of the system taken as unity. Note the eccentric location of the Sun. (Reprinted with permission from *The Astrophysical Journal*.)

Einstein[9] stood at the center of attention. Shortly thereafter in September 1921 father presented his results at the annual meeting of the British Association for the Advancement of Science at Edinburgh. In November 1921 father further spelled out his theory at an informal gathering in Leiden, where, except for a few Dutch scholars, Einstein and James Jeans[10], the well-known English astrophysicist, were present.[11]

Briefly the theory suggested that stellar orbits as we observe them are just a local phenomenon (see Figures 13 and 14). In reality, the movement of both 'streams' of stars is not straight-lined; rather both move in circular-formed rings but in opposite directions, thus against one another. This circular movement helps explain the flatness of the form of the stellar system that stems from the

---

[9] A. Einstein (1879–1955) published his 'special theory of relativity' in 1905 and his 'general theory of relativity' in 1915/16.

[10] The Englishman J. H. Jeans (1877–1946) was a gifted mathematician making major contributions in physics and later astrophysics and cosmology.

[11] Because Jeans was present at the Edinburgh meetings of the BAAS, it was known to him prior to publication. Working independently, though relying on much the same data, Jeans reached basically the same conclusions as Kapteyn. See Jeans to Kapteyn, 28 December 1921 (University of Groningen Archives), and Jeans, J. H.: 1922, 'The Motions of the Stars in a Kapteyn-Universe', *Royal Astronomical Society, Monthly Notices* 82, 122–32. For a discussion of Jeans's support of the 'Kapteyn Universe', with his force studies, and implications for the 'island universe' theory, see Smith, R.: 1982, *The Expanding Universe: Astronomy's 'Great Debate', 1900–1931*, Cambridge University Press, Cambridge, England, pp. 104–5.

observation of the number of stars, which would not be possible to explain without this movement. Although we do not observe the curvature of the circles, we understand the movement as straight-lines because the differences within a straight line are not visible. Indeed, only the stars that are our closest neighbors do we actually see move. While I cannot go deeper into this subject, it can be studied further in my father's own words in the May issue of the *Astrophysical Journal* – his last contribution to a foreign journal.[12]

My father's introduction to the structure of the universe which came to be called the 'Kapteyn Universe' – stands as a first exploration in a new area; it is also the closing chapter of his life's work.[13] It was the close of his life's work because the deadly illness, which would soon make an end of his work and struggle, already had its grip on him. He was tired and depressed and he could sit still for hours in his chair resting. No one suspected a terminal illness, least of all himself, because the doctors could find nothing of substance to report.

My parents decided to spend time with my sister – their daughter and son-in-law Professor Noordenbos – in Amsterdam from which they could easily reach Hilversum where they wanted to look for a house in which to spend their last years together. After looking awhile, they found a house that filled all their expectations. They visited it in the morning and approved it and that afternoon father brought my mother the keys to their own house. They were happy with their new home, a jewel in their eyes, offering rest and happiness hopefully for many years. Unfortunately they would not be allowed to live in it together, for his illness grew worse. For a half year the family lived between hope and fear because father's dear life slowly ebbed. Still, they had a half year of close unity that was so rich and happy in love that a few days before his death father could say with glowing tenderness: "Children, this is still the happiest time of my life." He bore his suffering like a hero; no one ever heard him complain, for he offered only uplifting and wise words. The large, sunny room where he lay for months, looked after by my mother, my sister, and me, we called 'the haven of delight' for it was truly a harbor of rest and happiness for whomever was feeling down. His cheerful face and hopeful plans for the future, his interesting stories and cheerful humor, his happy laugh and warming love – all these made the room a blessing. It became a consecrated place where people could not be anything except strengthened and encouraged and relieved of those burdens that slowly moved onward. Many came to visit him hoping against all hope that this good friend would be spared, for they loved him with a love that is special in life.

In April 1922 the International Astronomical Union meetings were held in

---

[12] For a thorough technical discussion of Kapteyn's entire theory, see Paul, Robert E.: 1993, *The Milky Way Galaxy and Statistical Cosmology*, Chapter 6.

[13] Kapteyn's 1920/22 theory of the stellar system became the epitome of investigations for providing the classical solution to a *statistical* cosmology, and in the process came to be known as the 'Kapteyn Universe', a term coined by Sir James Jeans. See Paul, Robert E.: 1993, *The Milky Way Galaxy and Statistical Cosmology, 1890–1924*, Chapter 6.

Rome. Because father was unable to attend, supported by pillows he wrote-out a list of activities for the coming years for which he desired the cooperation of astronomers. His old strength blazed on and this list – his scientific last testament – made a great impression when Willem de Sitter read it at the meetings. Everyone wanted to cooperate; thus it became a success for father. Here it was that Jules Baillaud, the French chairman of the meetings, said at the beginning of the meetings that, along with two other factors, the great progress of astronomy may be attributed in the last half century to the Groningen Laboratory.[14] A few days later when father heard that a lecture in England had been entitled 'The Kapteyn Universe' he smiled happily.[15] His life's greatest wish had been filled: He had served science and could restfully lay down his head. But the stressful work had tired and weakened him. His body could no longer contain his spirit. On the 24th of May he wrote his last letter. It was a greeting to the Royal Astronomical Society, which was celebrating its 100th anniversary.

On the 18th of June came the end. There was a holy quiet, a pain that hurt so much for his family that death was a blessing. Isn't death a blessing if it can bring such noble rest? Many trusted friends followed him to the cemetery at Westerfeld. There were no official tributes, for which his family was thankful, that interfered with the spirit; just true love and friendship proceeded and followed him. Peter van Anrooy played on the organ the gripping closing choir of the Matthaeus-Passion from Bach:

> Wir setzen uns mit Tranen nieder
> Und rufen dir im Grabe zu:
> Ruhe sanfte, sanfte Ruh.

His friend Bordewijk spoke a few loving words with a trembling voice. That was everything. The beautiful music of friends, their deep-feeling words, and the quiet pain that all felt at that hour of holy dedication. That evening in Groningen the director of the Harmonie-Orkest, Kor Kuiler, performed the death march by Beethoven as a memorial to the great Groninger.

And over the entire world wherever he was known, father was mourned. Everyone felt that one of the great minds of humanity, one who was great in spirit and courage who would be sorely missed, had departed. But they also knew that his spirit would live on and that they would build upon his work.

---

[14] See *supra*, p. 38.
[15] See Jeans, J. H.: 1922, 'The Motions of the Stars in a Kapteyn-Universe', *Royal Astronomical Society, Monthly Notices* **82**, pp. 122–32.

REFERENCES

I. Kapteyn Biographies:

Blaauw, Adriaan: 1973, 'Kapteyn, Jacobus Cornelius', *Dictionary of Scientific Biography* **7**, 235–40.
de Sitter, Willem: 1923, 'Nekrologe – Jacobus Cornelius Kapteyn', *Vierteljahrsschrift der Astronomische Gesellschaft* **lviii**, 162–90.
de Sitter, Willem: 1932, 'Further Observational Advances. The Survey of the "Local System'. The Life-Work of J. C. Kapteyn', in *Kosmos*, Harvard University Press, Cambridge, pp. 52–77.
Eddington, Arthur S.: 1922, 'J. C. Kapteyn', *The Observatory* **45**, 261–65.
Eddington, Arthur S.: 1923, 'Jacobus Cornelius Kapteyn, 1851–1922', *Royal Society of London, Proceedings*, (Section A) **cii**, xxix-xxxv.
Jackson, J.: 1922, 'J. C. Kapteyn', *Royal Astronomical Society, Monthly Notices* **83**, 250–55.
Kobold, Hermann: 1922 'J. C. Kapteyn', *Astronomisches Nachtrichen* **216**, 143–44.
Pannekoek, Anton: 1922, 'J. C. Kapteyn und sein astronomische Werk', *Die Naturwissenschaften* **10** (45), 967–80.
Paul, E. Robert: 1994, 'J. C. Kapteyn', in J. Lankford (ed.), *Encyclopedia for the History of Astronomy*, Garland Publ., New York.
Seares, Frederick H.: 1922, 'J. C. Kapteyn', *Astronomical Society of the Pacific, Publications* **34**, 233–53.
Strömgren, S. E.: 1922, 'J. C. Kapteyn', *Nordisk Astronomisk Tidskrift. Kobenhavn* **3**, 123.
Tenn, Joseph S.: 1991 (September/October), 'Jacobus Cornelius Kapteyn: The Tenth Bruce Medalist', *Mercury* **xx** (5), 145–47, 159.
van Maanen, Adriaan: 1922, 'Kapteyn Obituary', *Astrophysical Journal* **56**, 145–53.
van Rhijn, Pieter J.: 1922, 'J. C. Kapteyn', *Popular Astronomy* **30**, 628–32.

II. Secondary Works:

Berendzen, Richard, Hart, Richard and Seeley, Daniel: 1976, *Man Discovers the Galaxies*, Science History Publications, New York.
Bok, Bart: 1977, 'The Universe Today', in D.W. Corson (ed.), *Man's Place in the Universe: Changing Concepts*, University of Arizona Press, Tuscon.
Gingerich, Owen (ed.): 1984, *Astrophysics and 20th Century Astronomy to 1950, Part A*, Cambridge University Press, Cambridge.
Herrmann, Dieter: 1984, *The History of Astronomy from Herschel to Hertzsprung*, Cambridge University Press, New York.
Hoskin, Michael A.: 1963, *William Herschel and the Construction of the Heavens*, W.W. Norton & Co., Inc., New York.
Hoskin, Michael A.: 1982, *Stellar Astronomy: Historical Studies*, England: Science History Publ., Bucks.
Kevles, Daniel J.: 1971, '"Into Hostile Political Camps": The Reorganization of International Science in World War I', *ISIS* **62**, 47–60.
Paul, E. Robert: 1976, *Seeliger, Kapteyn and the Rise of Statistical Astronomy* (unpublished dissertation), Bloomington, Indiana, 473 pp.
Paul, E. Robert: 1978, 'The Nature of the Nebulae', *Journal for the History of Astronomy* **9** (3), 222–24.
Paul, E. Robert: 1981, 'The Death of a Research Programme: Kapteyn and the Dutch Astronomical Community', *Journal for the History of Astronomy* **12** (2), 77–94.
Paul, E. Robert: 1982, 'Festschrift for Oort', *Journal for the History of Astronomy* **13** (2), 141–42.
Paul, E. Robert: 1984, 'Kapteyn and Statistical Astronomy', in H. v. Woerden *et al.* (eds.), *The Milky Way Galaxy*, D. Reidel Publ., Dordrecht, Holland, pp. 25–42.

Paul, E. Robert: 1986, 'J. C. Kapteyn and the Early Twentieth-Century Universe', *Journal for the History of Astronomy* **17** (3), 155–82.

Paul, E. Robert: 1993, *The Milky Way Galaxy and Statistical Cosmology, 1890–1924*, Cambridge University Press, New York.

Paul, E. Robert: 1994, 'Stellar Statistics', in J. Lankford (ed.), *Encyclopedia for the History of Astronomy*, Garland Publ., New York.

Schuller tot Peursum-Meijer, J.: 1983, 'De sterrenkunde voor Kapteyn (1614–1878)', in A. Blaauw *et al.*, *Sterrenkijken Bekeken*, Groningen, 7–31.

Seeley, Daniel and Berendzen, Richard: 1972, 'The Development of Research in Interstellar Absorption, c. 1900–1930', *Journal for the History of Astronomy* **3** (1), 52–64, **3** (2), 75–86.

Seeley, Daniel and Berendzen, Richard: 1976, 'Astronomy's Great Debate', *Mercury* **7**, 67–71, 88.

Smith, Robert W.: 1982, *The Expanding Universe: Astronomy's 'Great Debate', 1900–1931*, Cambridge University Press, Cambridge, England.

Thoren, Victor E., Gow, Charles and Honeycutt, Kent: 1974, 'An Early View of Galactic Rotation', *Centaurus* **18**, 301–14.

van de Kamp, Peter: 1965, 'The Galactocentric Revolution, a Reminiscent Narrative', *Astronomical Society of the Pacific, Publications* **77**, 325 (325–35).

van Woerden, Hugo, Brouw, W. N. and van de Hulst, H. C. (eds.): 1980, *Oort and the Universe*, D. Reidel Publ., Dordrecht.

Willink, Bastiaan: 1991, 'Origins of the Second Golden Age of Dutch Science after 1860: Intended and Unintended Consequences of Educational Reform', *Social Studies of Science* **21**, 503–26.

Wright, Helen: 1966, *Explorer of the Universe: A Biography of George Ellery Hale*, Dutton & Co., New York.

Wright, Helen: 1975, 'Hale, George Ellery', in C. C. Gillispie (ed.), *Dictionary of Scientific Biography*, Charles Scribner's & Sons, New York, 16 vols. **4**, pp. 26–34.